Galaxy Morphology

AAS AMERICAN ASTRONOMICAL SOCIETY | IOP | ebooks™

AAS Editor in Chief

Ethan Vishniac, Johns Hopkins University, Maryland, USA

About the program:

AAS-IOP Astronomy ebooks is the official book program of the American Astronomical Society (AAS) and aims to share in depth the most fascinating areas of astronomy, astrophysics, solar physics, and planetary science. The program includes publications in the following topics:

Books in the program range in level from short introductory texts on fast-moving areas, graduate and upper-level undergraduate textbooks, research monographs, and practical handbooks.

For a complete list of published and forthcoming titles, please visit iopscience.org/books/aas.

About the American Astronomical Society

The American Astronomical Society (aas.org), established 1899, is the major organization of professional astronomers in North America. The membership (~7,000) also includes physicists, mathematicians, geologists, engineers, and others whose research interests lie within the broad spectrum of subjects now comprising the contemporary astronomical sciences. The mission of the Society is to enhance and share humanity's scientific understanding of the universe.

Editorial Advisory Board

Galaxy Morphology

B W Holwerda
Department of Physics and Astronomy,
University of Louisville, Louisville, KY, USA

IOP Publishing, Bristol, UK

ISBN 978-0-7503-3499-0 (ebook)
ISBN 978-0-7503-3497-6 (print)
ISBN 978-0-7503-3500-3 (myPrint)
ISBN 978-0-7503-3498-3 (mobi)

DOI 10.1088/2514-3433/ac2c7d

Version: 20220101

AAS–IOP Astronomy
ISSN 2514-3433 (online)
ISSN 2515-141X (print)

British Library Cataloguing-in-Publication Data: A catalogue record for this book is available from the British Library.

Published by IOP Publishing, wholly owned by The Institute of Physics, London

IOP Publishing, Temple Circus, Temple Way, Bristol, BS1 6HG, UK

US Office: IOP Publishing, Inc., 190 North Independence Mall West, Suite 601, Philadelphia, PA 19106, USA

Supplementary material is available for this book from http://iopscience.iop.org/book/978-0-7503-3499-0.

Cover image: Majestic spiral galaxy UGC 2885. Image credit: NASA, ESA and B. Holwerda.

Acknowledgement: This publication uses data generated via the Zooniverse.org platform, development of which is funded by generous support, including a Global Impact Award from Google, and by a grant from the Alfred P. Sloan Foundation.

For my wife Robin Allen,
our children Charlotte and Marten,
and in memory of Ron Allen and Mars Holwerda

Contents

Preface

This book is aimed at students of extragalactic astronomy at any level. It aims to bring together all the different ways astronomy as a science tried to quantify the appearances of galaxies.

Each topic is introduced and explained but that does not mean there is not much more to learn about each topic. I have tried to include a further reading section in each chapter. But science never sits still. I can assure you that as you read this, one of the topics has a new and interesting paper with new ideas that came out since I wrote this sentence.

Acknowledgement

Books happen because many different people help along the way. First, bravo to the IOP team to planting the idea and then letting myself talk me in to it.

Second, I owe a debt of gratitude to my family who let me hide upstairs and bang most of this out during the COVID quarantine. A big thank you to Governor Andy Bashear for keeping it together in Kentucky.

I got stuck on several occasions so I asked my friends and colleagues for help. Rob Crain: thank you for the references to get me started on the simulated galaxies section in Chapter 10. Lee Kelvin, thank you for talking about galaxy morphology with me on several occasions over the years, often as there just happen to be a beer in our hands. Pauline Barmby, for encouragement at just the right time and a Dean's perspective. Boris Haußler and Claudia Scarlata for organizing the "Galaxy Morphometrics" conference with me at the Lorentz center in 2016. Amanda Moffett for visiting and discussions on galaxies and morphologies. Johan Knapen for looking at a first draft. Chris Conselice for encouragement.

Preethi Nair for organizing the "Rings in Galaxies" conference at the University of Alabama in honor of Ron Buta. This amazing conference got me thinking more about other morphological features.

Heikki Salo for organizing the S4G meeting in Oulu Finland. I attended together with an 18 month old so my memory is sketchy but I think I learned a lot.

Bill Keel and Karen Masters are amazing Galaxy Zoo people and all-round awesome. Thank you for inspiring different morphological projects.

A big thank you to Piet van der Kruit for taking me on as a graduate student all these years ago and giving the opportunity to go to Baltimore and work with Ron Allen.

Author biography

B W Holwerda

Dr Holwerda is an associate professor at the University of Louisville. He has gotten involved in a variety of galaxy morphology studies in a large part because he originally had to pick out distant galaxies during his thesis. Things just sort of snowballed from there. Much of the initial writing happened during the COVID-19 quarantine. I would like to thank Robin Allen, Charlotte, and Marten for letting me get on with it.

Chapter 1

Introduction

Galaxy morphology is traditionally thought of how galaxies appear in their (stellar) light. Still the vast majority of descriptions of galaxy morphology are of images in wavelengths that can be attributed mostly to their stars.

The final shape and appearance of galaxies is the end-product of everything that has happened to these "island universes" to the point of observation. In principle, every evolutionary process has left its fingerprints. Describing galaxy morphology quantitatively is therefore a way to start understanding these processes that shape galaxies.

Why study Galaxy Morphology?

I do not go into the why in this book much. There is an excellent book by Chris Conselice that deals with much of the motivation for the study of galaxy morphology: to infer the rates of mergers, to discriminate between rotating and dispersion supported stellar systems, to figure out if the morphologies we see are in fact fundamental or not to the nature of galaxies. There (and many others) are valid reasons. I assume you have yours.

Morphology can be described at different scales: galaxy-wide (Chapter 2) at the scales of stellar nurseries, the groupings of stars that are born at approximately the same time (Section 4.9) and structures in between (e.g., Chapter 4), which hint at wide-spread phenomena that shape galaxy appearance. In principle, one could describe galaxy morphology down to the resolution of individual stars, their type, orbital parameters etc. Some of that information is now known for parts of the Milky Way and Andromeda, but for morphology, we limit our scales to that of starcluster on the smallest and the galaxy including the faint surrounding as a whole on the largest scales.

doi:10.1088/2514-3433/ac2c7dch1

In this book, I will go over historical classification schemes, descriptions of the galaxy light on a galaxy-wide scale, the variety of galaxy substructures, modern visual classifications, galaxy morphometrics, the low surface brightness features surrounding the classical galaxy, and the different kinds of star clusters.

1.1 Earlier Classifications

There are many different classifications of galaxies that were done on the first photographic plates, sometimes before we had a good idea what these fuzzy objects were. A few of these early classes are still around. For example, the "cD" class galaxy (Morgan 1958), mostly for historical reasons. De Vaucouleurs' reference catalog (de Vaucouleurs et al. 1991) included several of such classification. What these all had in common was that the original classification was done (a) by eye and (b) by a single experienced astronomer. Only one of these classification schemes survives to this day and we owe a lot of our current nomenclature to it: Hubble's Tuning Fork.

1.2 Hubble's Tuning Fork

Edwin Hubble classified the "nebulae" as they were called at the time into the now iconic tuning fork (Hubble 1926, 1936). His scheme is widely adopted and updated (Sandage 1961; de Vaucouleurs 1959; Sandage 1975). Figure 1.1 shows the modern schematic, Figure 1.2 the original description. Broadly there are two classes: ellipticals (left) and disk or spiral galaxies (right). That bimodaly has held up well over time. Within spiral galaxies, there are those with a bar, an elongated structure in the middle, and those without. Subclass of spiral is how prominent the central bulge of stars is and how tightly wound the spiral arms. Subclasses within ellipticals is how flattened of a football is their appearance (from soccer ball to a deflated American football). The central idea is that one gradually goes from perfectly round spheroidal[1] to a flattened disk. Very logical! Wrong, but logical.

Early/Late Type Galaxies

Early interpretation of Hubble's schematic was that galaxies start spheroidal and end up flattened. So the nomenclature has unfortunately inherited some of this interpretation: ellipticals are referred to as "early type" and disk galaxies as "late type."

Add to this, that we now know that two spirals crashing into one another often end up making a spheroidal. So elliptical galaxies are "early types" that often formed from the merger of two "late types." With me so far?

It is legacy nomenclature and it will be encountered in the literature but in future writing, one can stick to better descriptions: elliptical or spheroidal instead of "early type" and disk or spiral galaxy instead of "late type."

[1] Fancy word for blob. I prefer blob.

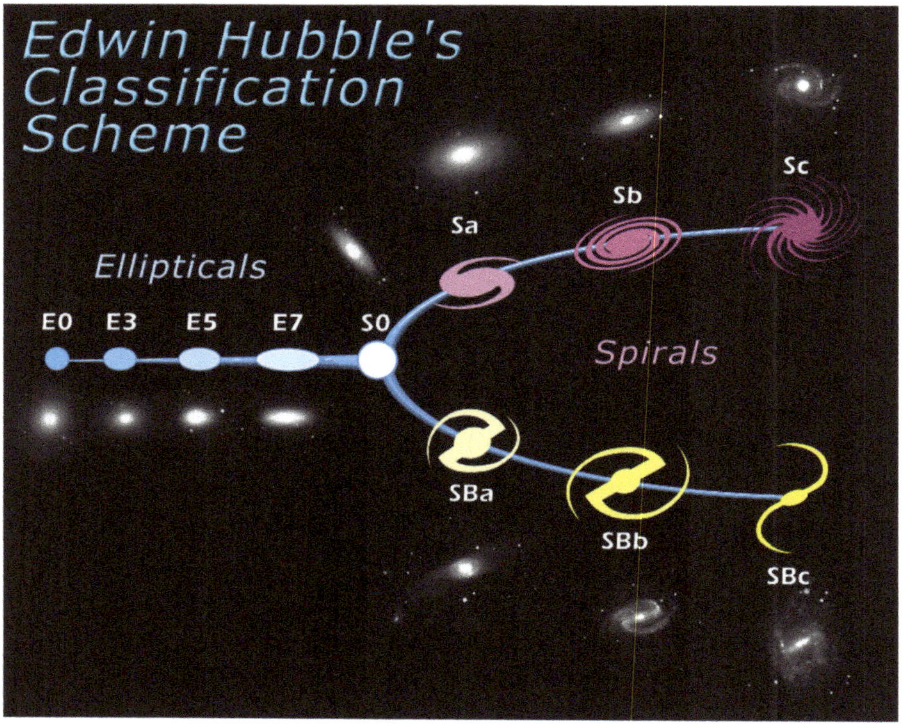

Figure 1.1. The Hubble Tuning Fork from https://www.spacetelescope.org/ a visual representation of the classification scheme put forward by Edwin Hubble. Figure reprinted with permission from NASA & ESA/ Hubble.

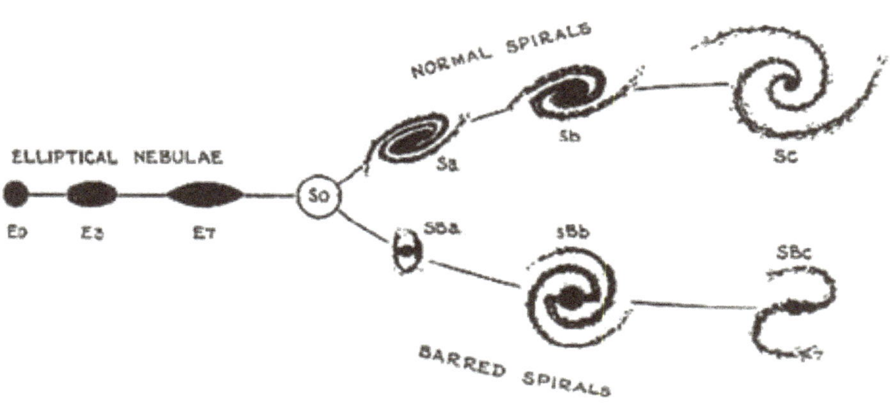

Figure 1.2. The original Hubble Tuning Fork (originally Hubble 1936, reproduced in Buta 2011). Note the use of "nebulae" and "normal" versus "barred" spiral. S0 is almost a missing link in the schematic. Figure reproduced from Buta (2011) Copyright © 2013, Springer Science Business Media Dordrecht. With permission of Springer.

Elliptical subclassifications are among the visible flattening, ranging from E0 (ellipticity=0) to E7 (ellipticity = axis ratio=7). The spiral subclassifications range from Sa to Sc to denote the diminishing prominence of the bulge, and get a capital "B" if they have a bar (e.g., SBc). The class that ties it all together is S0, what looks like an elliptical but with a bit of fringe of a disk.

Fantastic! We are done! Classify everything like an ornithologist and patterns and evolutionary tracks are bound to appear. Except there are two categories that the tuning fork ignores: irregulars and peculiars. Both are catch-all bins of everything that cannot be classified easily: amorphous dwarf galaxies without a clear central bulge, train-wreck galaxy–galaxy collisions and a range of other, often rarer, galaxy appearances. The classification looks clear and seemed to bring order but only because there is a "misc" folder lurking offstage.

The irregular galaxy class is the most common occurring galaxy once you correct for the missing ones because they're intrinsically faint. Peculiars are the various stages of a merger between galaxies. Rare but the fact that we can see them at all means it must happens regularly.[2]

Another issue with this classification is that it says almost nothing useful about the ellipticals. If one starts with an American football,[3] one can go from E7 to E0 simply by rotating it: the elliptical classification is for a large part viewing angle! That is not a characteristic of the object but how we look at it.

Therefore there have been attempts to classify ellipticals according to some actual observed property: the amount of rotation (Cappellari et al. 2011), outer isophote (Kormendy & Bender 1996), or the light profile (see Chapter 2). This is how there are sub-categories within ellipticals that are defined by a specific (actually inherent) property: E+A galaxies look elliptical but are recently past a burst of star formation.

That is not to say that "classical" visual morphological classification is no longer in use or used! Modern classifications of digital images of resolved galaxies, of which we have increasing amounts, is incredibly useful for classification algorithms. These "classical" morphology classification efforts (Nair & Abraham 2010a, 2010b) are diligently being improved from thousands of galaxies (Baillard et al. 2011), to tens of thousands (Nair & Abraham 2010a). Such expertly classified samples are ideal training sets for machine learning (see Chapter 11) especially if one holds back a fraction for a blind classification test ("closed envelope" test).

Different Scales

I want you to note at this point that galaxy classification has used to scales to differentiate them: galaxy-wide appearance to classify into broad categories (elliptical versus spiral[4]) and then often sub-galaxy features, such as bulge or bar strength, to sub-classify.

[2] See the book "The Cosmic Evolution of Galaxy Structure" by Chris Conselice (2020) on their importance.
[3] Or rugby, that should visualize if for much of the rest of the world.
[4] Not much of that early/late type shenanigans in the rest of this book if I can help it.

1.3 The 90% Rule

When things in studies of galaxies are stated in a definitive tone, please remember that this is a phenomelogical description of very large populations of objects now. If we say things like "Elliptical galaxies are red," then this is true for some 80%–90% of them. You can find blue ones (Moffett et al. 2019). Or that "spiral galaxies have blue arms." There are red spiral galaxies (Masters et al. 2010). This is not mathematics and categorical statements in this book or elsewhere should be taken as "'true for 90%+ of a galaxy sample" unless the confidence level is stated somewhere. This is important to remember. A counter-example does not invalidate the 90% truth but it points to an interesting sub-sample, or a short evolutionary phase.

1.4 Assignments

Accompanying this chapter is the "Hubble Tuning Fork" jupyter notebook, available at https://iopscience.iop.org/book/978-0-7503-3499-0.

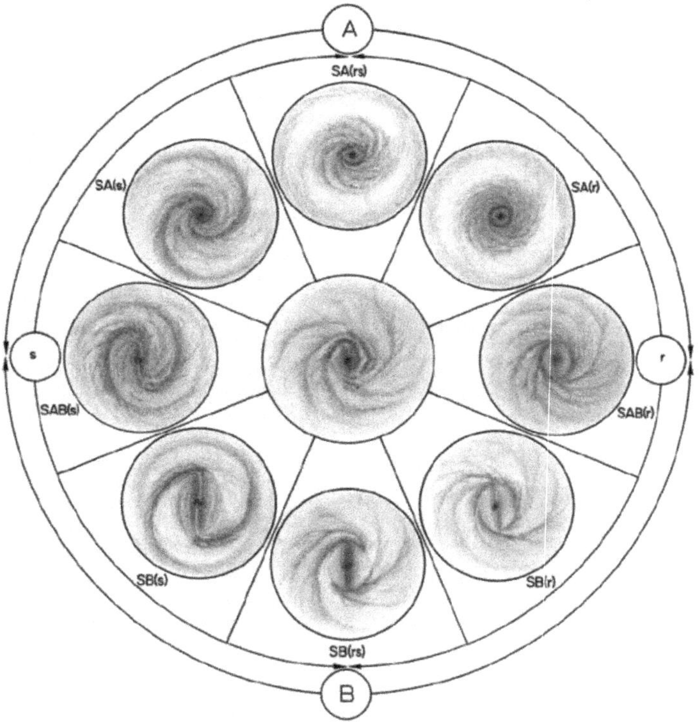

Figure 1.3. The classification aid for de Vaucouleurs for the Reference Catalog (RC3) to classify disk galaxies into different spiral galaxy subclasses. Original in the collection of R. Buta. Reprinted from Buta et al. (2007) © The Authors.

1.5 Further Reading

The overview by Ron Buta (2011) is excellent and goes into great detail in visual morphology classification. The actual Reference Catalog (de Vaucouleurs et al. 1991; see Figure 1.3) or Uppsala General Catalog (UGC; Nilson 1973) also describe these classifications. Visual galaxy classifications are still very much at the forefront of astronomical science, see for example Nair & Abraham (2010a).

References

Baillard, A., Bertin, E., de Lapparent, V., et al. 2011, A&A, 532, A74

Buta, R. J. 2011, arXiv:1102.0550

Buta, R., Corwin, H. G., & Odewahn, S. C. 2007, The de Vaucouleurs Atlas of Galaxies (Cambridge: Cambridge Univ. Press)

Cappellari, M., Emsellem, E., Krajnović, D., et al. 2011, MNRAS, 416, 1680

Conselice, C. J. 2020, The Cosmic Evolution of Galaxy Structure (Bristol: IOP Publishing)

de Vaucouleurs, G. 1959, AJ, 64, 397

de Vaucouleurs, G., de Vaucouleurs, A., Corwin, H. G., et al. 1991, Third Reference Catalogue of Bright Galaxies, Vol. 1–3, XII (Berlin: Springer-Verlag)

Hubble, E. P. 1926, ApJ, 64, 321

Hubble, E. P. 1936, Realm of the Nebulae (New Haven, CT: Yale Univ. Press)

Kormendy, J., & Bender, R. 1996, ApJL, 464, L119

Masters, K. L., Mosleh, M., Romer, A. K., et al. 2010, MNRAS, 405, 783

Moffett, A. J., Phillipps, S., Robotham, A. S. G., et al. 2019, MNRAS, 489, 2830

Morgan, W. W. 1958, PASP, 70, 364

Nair, P. B., & Abraham, R. G. 2010a, ApJS, 186, 427

Nair, P. B., & Abraham, R. G. 2010b, ApJL, 714, L260

Nilson, P. 1973, Uppsala General Catalogue of Galaxies (Uppsala: Astronomiska Observatorium)

Sandage, A. 1961, The Hubble Atlas of Galaxies (Washington, DC: Carnegie Institution)

Sandage, A. 1975, PASP, 87, 853

AAS | IOP Astronomy

Galaxy Morphology

B W Holwerda

Chapter 2

Galaxy Scale

2.1 Introduction

The first descriptions we will review are galaxy-wide descriptions of the light distribution of galaxies. As it was noted earlier, galaxies are (mostly) pretty symmetric around at least one axis. So they lend themselves very well to polar coordinates: radius from the galactic center (R) and height above (or below) the plane (z) defined by the symmetry axis. Let us note here that these are some very basic assumption that we will use throughout: there is a center to a galaxy and a plane whose polar coordinates describe the distribution of starlight in this galaxy well. This is such a basic assumption that we call all galaxies that do not comply with it "irregular."[1]

2.2 de Vaucouleurs Profile

The de Vaucouleurs profile was defined to describe the spheroidal galaxies observed (de Vaucouleurs 1948, 1959). Because their elliptical type classification is dominated by our viewing angle, there really is not much of an obvious plane to define with these galaxies. The de Vaucouleurs profile is therefore described only in distance from the center of the galaxy in terms of natural logarithms:

$$\ln(I(R)) = \ln(I_0) - kR^{1/4}, \tag{2.1}$$

where ln is the natural logarithm, I_0 is the central surface brightness and k is a constant. More commonly, this is re-written to include the half-light radius (see also Chapter 3) since it gives a typical size of the galaxy.

[1] The most commonly occurring galaxies in the universe are irregulars. But because we saw the bright disk and ellipticals first. There is a lesson in observational biases here.

doi:10.1088/2514-3433/ac2c7dch2

$$\ln(I(R)) = \ln(I_e) + 7.669\left(1 - \left(\frac{R}{R_e}\right)^{1/4}\right) \tag{2.2}$$

where I_e is the surface brightness at the effective radius. Radii are typically in on-sky units (arcseconds or arcminutes) until there is a good distance to the galaxy and it can be expressed in physical units.

The de Vaucouleurs profile is more generalized to the Sérsic profile (see below). Because this relations always seemed to hold for elliptical galaxies, it is sometimes referred to as "'de Vaucouleurs' Law."[2] It was also noted that the bulges in spirals also typically exhibit such a profile. This led to the thinking that bulges in spirals are really just small elliptical galaxies surrounded by an additional disk (see Chapter 4).

2.3 Exponential Profile

To describe disk galaxies, an exponential radial profile describes the surface brightness best:

$$I(R) = I_0 e^{(-R/h)} \tag{2.3}$$

where I_0 is the central surface brightness and h is the scale length of the disk. This scale length can also be related to the half-light radius of the disk (see Chapter 3).

2.3.1 Freeman's "Law"

In the 1970s, Freeman noted that the central surface brightness (μ_0, which corresponds to I_0 in surface brightness) in disk galaxies is always the same and colloquially is known as "Freeman's Law."[3] Yet this was a selection effect. Low surface brightness disks were easily missed. Figure 2.1 shows the "law" and the central surface brightnesses of disks in a complete sample in a representative volume; there are central surface brightness values of every kind up to a limit where no brighter disks exist (McGaugh & Bothun 1994). Why did I introduce you to this law? To make you aware that selection effects lurk everywhere in observational astronomy (cf Figure 2.2 which shows the limits of the survey).

However, disk galaxies *always* have a central source, be it a bulge, a pseudo-bulge or a nuclear star cluster, there is always extra light "on top of" the disk profile. A variant of Freeman's Law is that the maximum surface density of stellar systems never seems to exceed $\Sigma^*_{max} = 3 \times 10^5 M_\odot/pc^2$ (Hopkins et al. 2010; Grudić et al. 2019), yet that also appears to be more of a selection effect with the maximum central stellar density depending on stellar mass (Ling et al. 2020). And how one should model the contribution of the disk to the central parts of a galaxy is still being reconsidered with newer additional information of spectroscopy for every place in a galaxy (see Figure 2.3), where the authors argue that for disk-dominated galaxies,

[2] Note that this is a good way to build in a bias. Calling it a law, makes it sounds immutable and never open to further observation. It is a good relation, describing most galaxies in this class reasonably well. No galaxy goes to jail for breaking it though.

[3] Yet another "law." There is a reason present-day astronomers are resistant to calling things a "law" now.

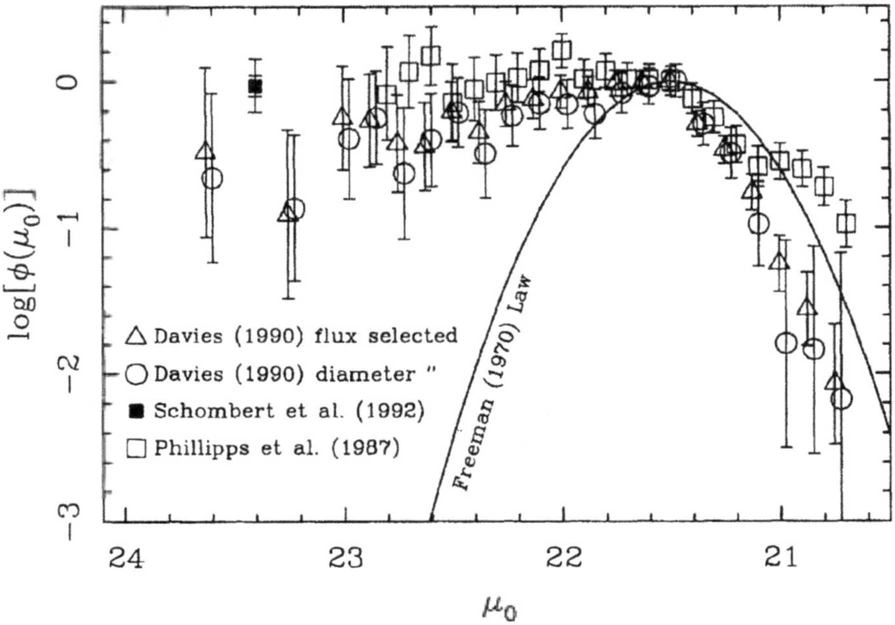

Figure 2.1. The central surface brightness distribution of disk galaxy fits for the local universe, corrected incompleteness. The combined data shows that low surface brightness galaxies dominate the numbers of disk galaxies and no "typical" central surface brightness can be identified. Figure reproduced from McGaugh et al. (1995). © 1995. The American Astronomical Society. All rights reserved.

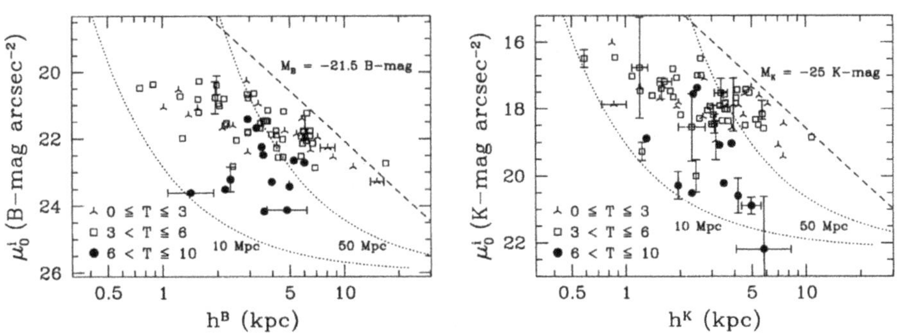

Figure 2.2. The central surface brightness measured in B- and K-band for different types of galaxies. Dashed lines denote the selection limits for different distances. Figure reproduced with permission from de Jong (1996b) © ESO.

the disk profile in the center may be flattened (green line) rather than strictly exponential.

2.4 Sérsic Profile

The Sérsic profile is the generalization of the exponential and de Vaucouleurs profiles. Graham et al. (2005) and Graham & Driver (2005) extensively treat how to

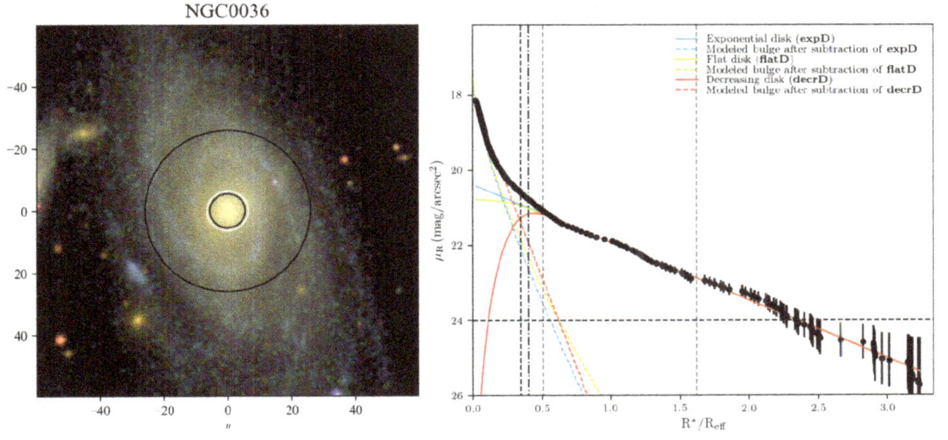

Figure 2.3. Left: SDSS color image of NGC 0036. The white circle overlaid in the SDSS image depicts the bulge diameter (with radius R_B) and the innermost black circle illustrates the bulge radius (R_C). The radius of the outermost black circle corresponds to the limit of the disk (R_D). Right: SDSS g-band SBP illustrating the assumed three disk configurations: observed profile (black points), an exponential disk (blue solid line) inwardly flattening disk (green solid line), and a centrally depressed disk (red solid line). Dashed lines are the bulge profile after subtraction of the respective disk model. Figure from reproduced with permission Breda et al. (2020) © ESO.

compare this profile and relate it to effective radius (see also Chapter 3) and photometry.

Sérsic's R1/n model (Sérsic 1963, 1968) is most commonly expressed as an intensity profile, such that:

$$I(R) = I_e \exp\left(-b_n\left[\left(\frac{R}{R_e}\right)^{1/n} - 1\right]\right) \tag{2.4}$$

where I_e is the intensity at the effective radius (R_e), which encloses half of the total light (see also Caon et al. 1993, and Ciotti 1991). The constant b_n is defined in terms of the third and final parameter n, which describes the "shape" or slope of the light profile. This constant (which depends on the value of n) can be generally computed using gamma functions:

$$\Gamma(2n) = 2\gamma(2n, b_n) \tag{2.5}$$

where Γ and γ are the complete and incomplete gamma functions respectively. For a de Vaucouleurs profile $b_{n=4} = 7.669$ and an exponential $b_{n=1} = 1.678$. In practice you will let the fit code deal with this (see below).[4] One can approximate this with:

$$b_n \simeq 2n - 1/3. \tag{2.6}$$

[4] Unless you suddenly are convinced you can code it "better" at which point I say "Godspeed."

The expression of intensity in Equation (2.4) can also be expressed as a surface brightness profile:

$$\mu(R) = \mu_e + \frac{2.5b_n}{\ln(10)}[(R/R_e)^{1/n} - 1] \tag{2.7}$$

where μ_e is the surface brightness at R_e. To get the central surface brightness ($R = 0$), one uses:

$$\mu_0 = \mu_e - \frac{2.5b_n}{\ln(10)}. \tag{2.8}$$

This central surface brightness was initially thought to be constant ("Freeman's Law," see Section 2.3.1 earlier) but also how the disk continues toward the center is still under debate (Figure 2.3).

The Sérsic profile is a generalized version that encompasses the de Vaucouleurs and the exponential profiles. If one sets $n = 4$ in Equation (2.4), one gets something identical to Equation (2.2). Similarly, if one sets $n = 1$, the exponential profile of Equation (2.3). One can then relate the scale length and central intensity of the exponential profile (h,I_0) to the effective radius (R_e) and intensity at the effective radius (I_e).

2.4.1 Sérsic Photometry

A big incentive to parameterize the light from galaxies in these profiles is to account for any light that falls outsize an aperture placed on the image. Stars are effectively point sources and much easier to account for (or at least uniformly accounted for). With all these different profiles how much light is one missing if the light is measured in an aperture (e.g., the circle in Figure 2.3)?

To estimate the enclosed luminosity of a galaxy within radius R, we integrate Equation (2.4):

$$L(<R_{\rm aper}) = \int_0^{R_{\rm aper}} I(R)2\pi R dR = I_e R_e^2 2\pi \frac{e^{b_n}}{(b_n)^{2n}} \gamma(2n, x) \tag{2.9}$$

where $x = b_n(R/R_e)^{1/n}$ However, for accurate photometry, one needs to include a correction for light lost due to the telescope's scatter (Jarrett et al. 2007). To infer the *total* luminosity, one can integrate to infinity. It all depends to what level you care. Typical corrections to infinity from an aperture (the galaxy is infinitely big!) are on the order of 10%.

2.5 Two-component Galaxies

Early CCD observations of galaxies allowed astronomers to accurately map the isophotes of galaxies (Figure 2.4). This meant that a 2D profile can be fit to light distribution and one can opt to model both the disk and the bulge with a profile of

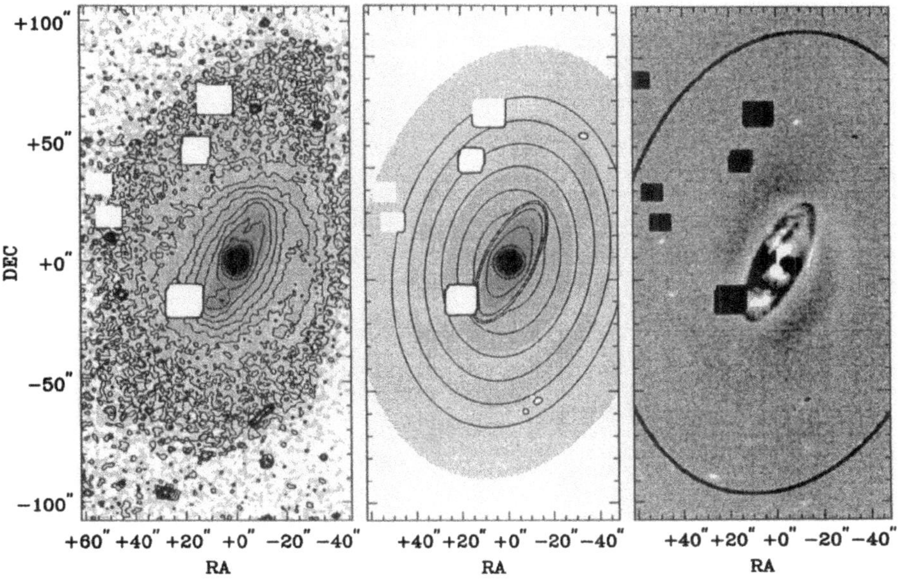

Figure 2.4. Two-dimensional fit to a galaxy. Left: the data with isophotes (equal light contours) with foreground stars masked out, middle: the two-dimensional model for this data, right: the residual with the model subtracted from the data. Figure reproduced with permission from de Jong (1996a) © ESO.

choice (e.g., both with a Sérsic profile or an exponential for the disk and a de Vaucouleurs one for the disk).

The early CCD studies showed how disk profiles behaved and one could not quantify the bulge-to-disk (B/D) or bulge-to-total (B/T) values. This value looked like a more objective way to quantify the galaxy's type (which even converted to a numerical value is an eyeball estimate), but the value changed with color (Figure 2.5). There were after all color gradients in galaxies and especially disk galaxies (de Jong & van der Kruit 1994; Courteau 1996; de Jong 1996a, 1996c;

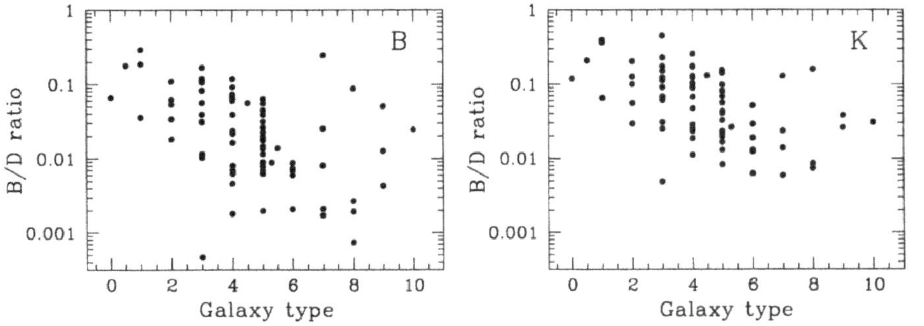

Figure 2.5. The bulge-to-disk ratio for different galaxy numerical types. Note how numerical prominence of the bulge correlates with visual type but it is a relation with a fair amount of scatter. Figure reproduced with permission from de Jong (1996b) © ESO.

Kassin et al. 2006a, 2006b). But in a near-infrared filter, where all stars emit light and the attenuating effects of dust are minimal, this seemed like a great way to morphologically quantify disk galaxies beyond the typing done by eye. In Section 2.10 later in the chapter, we go over the limitations of this approach.

2.6 Vertical Profiles

This was first proposed by van der Kruit (van der Kruit & Searle 1981) to map edge-on galaxies and started from this description of the three-dimensional distribution of light:

$$L(R, z) = L_0 e^{-R/h} \operatorname{sech}^2(z/z_0), \tag{2.10}$$

where R is the radius, z is the height above the plane, h and z_0 are the exponential scales in the radial and vertical direction. One can now approximate this in two cases, face-on and edge-on. In the face-on case, one integrates along the vertical dimension to obtain an expression for *surface brightness*:

$$\mu(R) = \mu_0 e^{-R/h} \tag{2.11}$$

which is the exponential profile we encountered earlier. And in the case of the edge-on expression:

$$\mu(R, z) = \mu(0, 0)\left(\frac{R}{h}\right) K_1(R/h) \operatorname{sech}^2(z/z_0) \tag{2.12}$$

where the position in the disk is still expressed in polar coordinates. The K_1 term is the modified Bessel function.[5]

To simplify fits to profiles (made averaging ellipse in the photometry) one can use approximations of these expressions for the edge-on case. The inner disk ($R/h \ll 1$):

$$\frac{R}{h} K_1\left(\frac{R}{2h}\right) \simeq 1 + \frac{R^2}{2} h^2 \ln\left(\frac{R}{2h}\right). \tag{2.13}$$

The outer disk ($R/h \gg 1$):

$$\frac{R}{h} K_1\left(\frac{R}{2h}\right) \simeq \sqrt{\frac{\pi R}{2h}} \exp\left(\frac{-R}{h}\right). \tag{2.14}$$

Close to the plane of the disk ($z/z_0 \ll 1$):

$$\operatorname{sech}^2\left(\frac{z}{z_0}\right) \simeq \exp(-z^2/z_0^2). \tag{2.15}$$

In the outskirts of the disk ($z/z_0 \gg 1$):

[5] This may look like a complication compared to the 3D expression but back 1981, a function with known look-up tables made this a lot easier to compute and therefore fit to actual data.

$$\text{sech}^2\left(\frac{z}{z_0}\right) \simeq 4\exp(-2z/z_0). \tag{2.16}$$

2.6.1 Vertical Profile Shapes

The sech^2 is a convenient expression for the vertical profile (van der Kruit & Searle 1981). It is "softened" in the plane to not peak as much as an exponential. This accounts for the inevitable smoothing that happens by the blurring effect of a telescope's optics (the point-spread function) and the fact that some of the light, no matter what the waveband is in which the galaxy is observed, will have attenuated in the plane of the disk thanks to interstellar dust. There is no physical motivation for this choice of vertical profile and sometimes sech (z) or another function is used.

The vertical profiles of galaxies, specifically those seen edge-on, show often evidence of a second component: there are two disks, a thin, high-surface brightness one and a thicker, low surface brightness one. See Chapter 7 for more on these low surface brightness thick disks.

2.6.2 The Effects of Inclination

So far, we have discussed ideal cases: perfectly face-on and edge-on. Realistically our viewing angle depends. Inclination is defined as the angle the normal on the disk makes with our line-of-sight (Figure 2.6).

Pohlen et al. (2007) explored how our viewing angle changes the measurement of the disk height and scale length. Figure 2.6 shows how much the scale length measurement depends on the actual inclination. Both dust in the plane and the inclination change the measures value for the scale length significantly.

2.6.3 Axes Ratio

There are three axes to consider when describing a galaxy, corresponding to the three-dimensional shape the stars occupy. Figure 2.7 illustrates the three kinds of shapes: spheroidal, prolate, and disks. In the case of elliptical galaxies, it is not immediately obvious if the galaxy is perfectly spheroidal or if it is a prolate galaxy seen along its major axis. The sub-classification along the Hubble sequence is therefore a convolution of our viewing angle and the actual shape. Elliptical and bulges span a range of shapes between prolate and perfectly spheroidal.

The trick now is to measure and fit a 3D shape of spheroidal (or just a disk or even prolate?) elliptical galaxies (cf Figure 7.32).

2.7 Sérsic Profile with Wavelength

A single Sérsic fit to a galaxy will depend in which wavelength the image is taken. Observations are typically taken in filters (wavebands) that cover a certain wavelength regime in the ultraviolet, optical, or infrared. Depending on the stellar population that dominates that waveband, the best describing Sérsic profile of

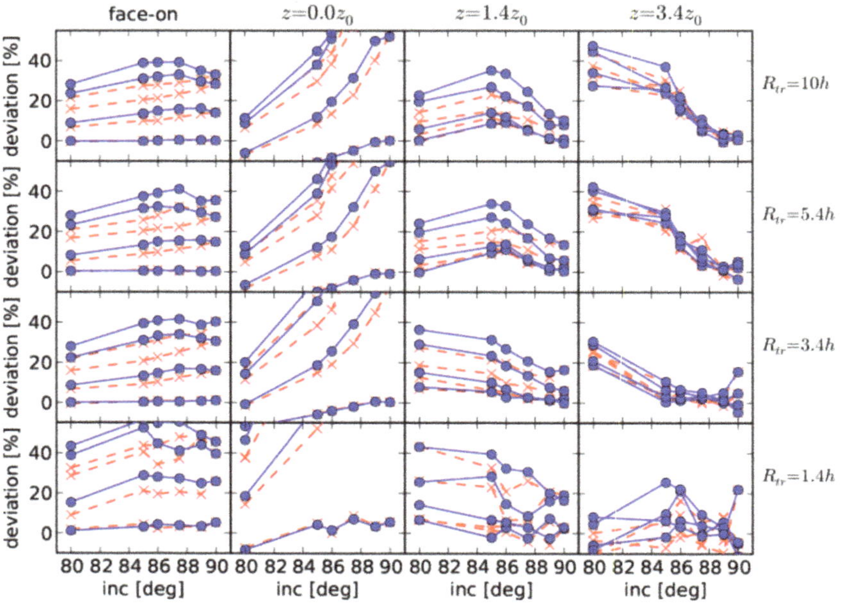

Figure 2.6. The systematic deviation of the measured scale length (h) with inclination and dust influence. Plotted is the deviation of the measured from the input scale length (in percentage) based on 96 mock images. From left to right the four columns are the results for the face-on-equivalent profiles, for the major axis, and for two cuts at 1.7x and 3.4x the scale height (z_0). The four rows correspond to disks truncated at different radii, from top to bottom: not truncated (mimicked by truncating at 10 × h), long truncated (truncated at 5.4h), intermediate truncated (truncated at 3.4h) and finally short truncated (truncated at 1.4h). The different symbols represent the two different ways to obtain an average quadrant: filled circles for the applied mean average out of four, crosses for the average out of two quadrants, excluding the dusty side. The different lines connecting the points in each panel are associated to the different dust distributions (ordered according to influence on the stellar disk). From bottom to top we have the transparent case, the best case, an intermediate case and finally the worst case. Figure reproduced from Pohlen et al. (2007). Copyright © 2007 The Royal Astronomical Society.

galaxies may change. Spheroidal galaxies are typically made up of only old stars so the profile shape (Sérsic index n) will stay constant, no matter what wavelength it is observed in. The light from disk galaxies however is dominated by young (blue) stars in the disk and older (red) stars in the central bulge. Therefore, as one steps through wavebands from blue to red, the same galaxy will appear more disk-like to bulge-like: the Sérsic profile will go from disk ($n < 2$) to more spheroidal ($n > 2$). Figure 2.8 shows the change of Sérsic profile (n) normalized to the blueish sdss-g filter: the galaxies with a low Sérsic index in the redder sdss-r band change their Sérsic index significantly going from blue to red (from disk to bulge dominated). The high Sérsic index galaxies ($n_r > 2.5$) remain close to the same profile in each waveband.

The effective radius meanwhile remains more constant for disk galaxies thanks to the trade-off from disk to bulge dominated (Figure 2.9). The steeper spheroidal galaxies shrink dramatically when going to redder wavelengths. Something to keep in mind when comparing galaxy sizes across redshifts! Correcting the observed

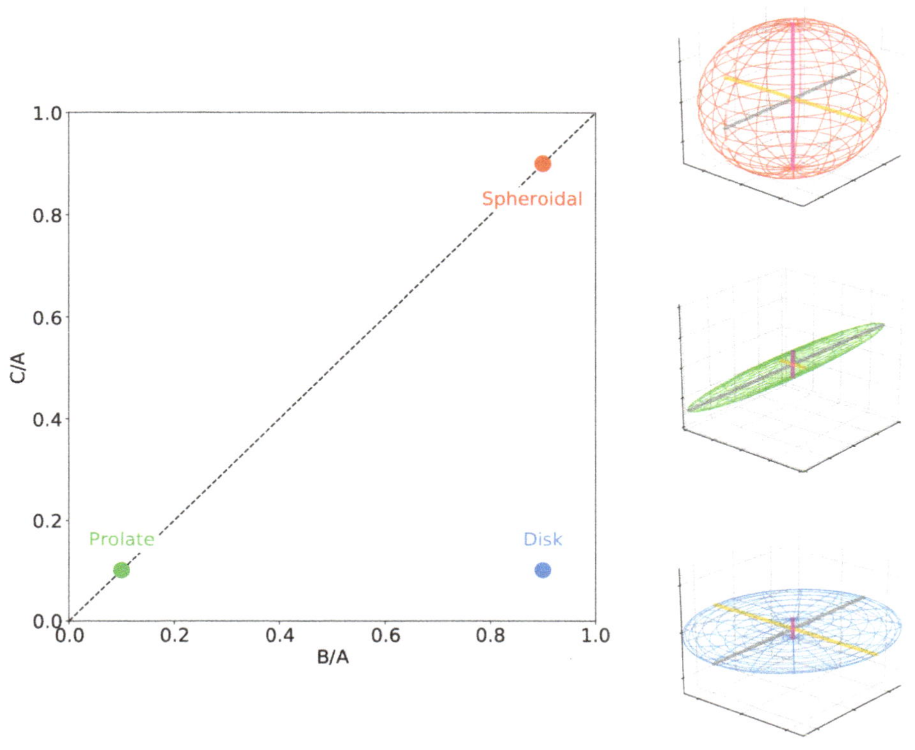

Figure 2.7. An illustration of the axes ratios of spheroidal, prolate, and disk galaxies. A, B, and C are the three axes describing the spheroidal in 3D. Spheroidal is when all three axes are similar in size, prolate when two are minor and one major, and a disk is when one axes is minor and the remaining two are of the same size. Elliptical galaxies could either be spheroidal or a more prolate shape. Figure reproduced from Kado-Fong et al. (2021).

wavelength for the effects of distance and hence redshift of the distant galaxy is called morphological K-correction (see Section 10.2.1).

UGC 2885 "Rubin's Galaxy," A Gentle Giant UGC 2885 is the largest disk galaxy in the nearby universe (nearest 100 Mpc). Originally noted by Vera Rubin as a disk of unusual size (a diameter of 250 kpc).[6] Quite how it managed to be 10× the diameter of the Milky Way without signs of major or minor interactions remains a bit of a mystery. In 2020, the galaxy was nicknamed "'Rubin's galaxy" in honor of the pioneering astronomer, see Figure 2.10.

[6] I like the GOUS—Galaxy Of Unusual Size (thank you Ethan Siegel) but I am told no one has seen "The Princess Bride" anymore so no one will get this reference.

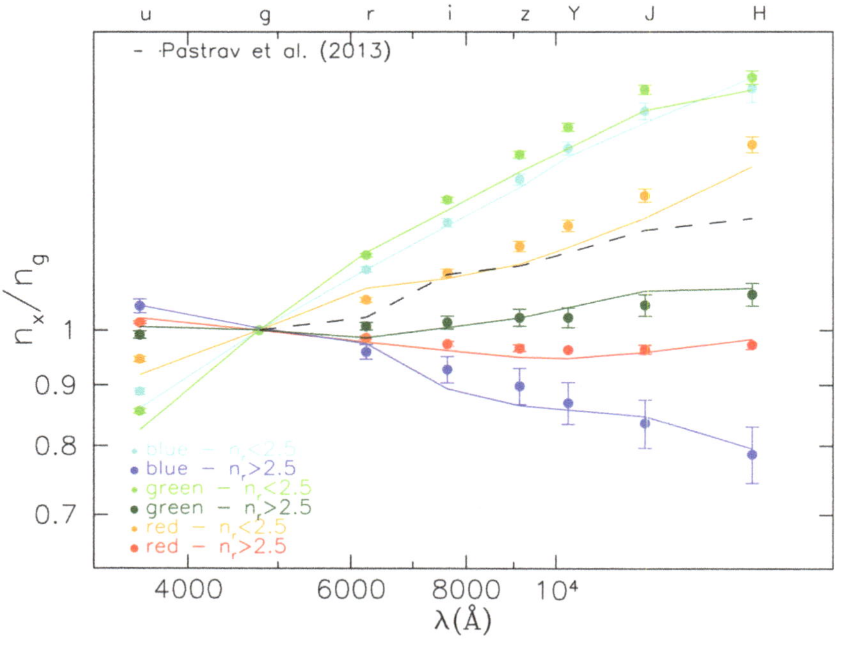

Figure 2.8. The median Sérsic index (n) as a function of waveband (from red to blue) from single profile Sérsic fits to GAMA data, normalized to sdss-g band. Galaxies are considered blue, green, and red according to their $u - r$ color and disky or spheroidal on their n_r Sérsic number. The spheroidals ($n_r > 2.5$) all show constant profile no matter what wavelength (all the same stellar population and hence color). The disk galaxies ($n_r < 2.5$) show a strong steepening toward redder colors: the bulge becomes dominant of the disk. Dashed line is the effect of dust on the profile predicted by Pastrav et al. (2013). Figure reproduced from Vulcani et al. (2014). Copyright © 2014 The Royal Astronomical Society.

2.8 Software Packages

The good news is that should you want to apply a Sérsic profile or an edge-on profile to a galaxy image you have obtained, you will not have to do so from scratch. There are several software packages out there that fit these descriptions to your image data. I will discuss them briefly here, so you can pick which one fits your needs best. Roughly, there are profile fits that are done as part of a general source detection and cataloging effort. These are often limited to a single component (no bulge+disk fits here) and little in the way of checks for accuracy.[7]

Catalog and Profile Packages

SOURCE EXTRACTOR (SE)—this package has been the workhorse for astronomical catalogs for the better part of three decades. Venerable does not mean out of date. It runs very quickly and in v3 allows one to specify a PSF file and extract profiles

[7] Most of these will provide a reduced χ^2 value but that sort of thing is only a good estimate of goodness-of-fit where the model is simple. Galaxies are far from it.

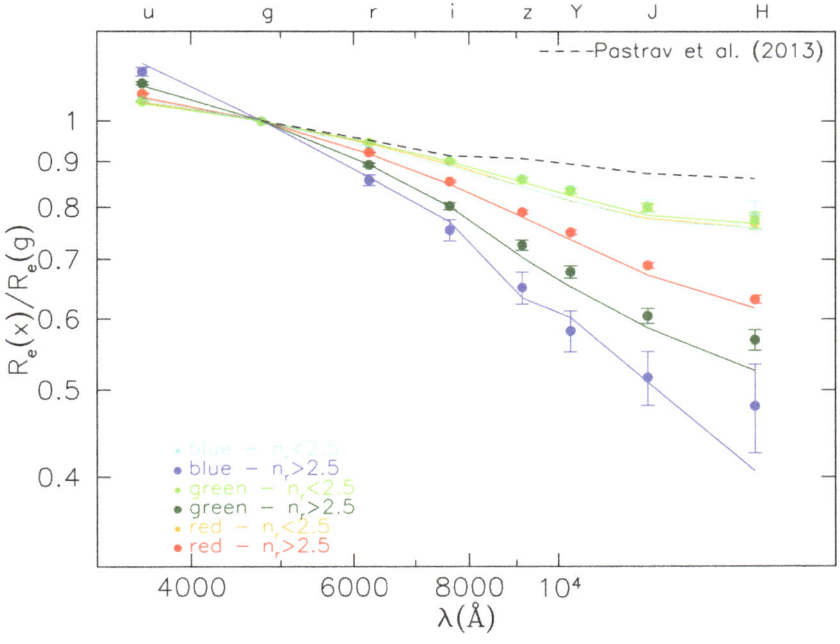

Figure 2.9. The median effective radius (r_e) as a function of waveband (from red to blue) from single profile Sérsic fits to GAMA data, normalized to sdss-g band effective radius ($R_e(g)$). Galaxies are considered blue, green and red according to their $u - r$ color and disky or spheroidal on their n_r Sérsic number. The spheroidals ($n_r > 2.5$) all show a strong dependence of size on wavelength. The disk galaxies ($n_r < 2.5$) show size more independent of color: the change from bulge to the disk-dominated light profile ensures 50% of the light is still at the same point. The dashed line is the effect of dust on the profile's effective radius predicted by Pastrav et al. (2013). Figure reproduced from Vulcani et al. (2014). Copyright © 2014 The Royal Astronomical Society.

(Sérsic, pure disk, or de Vaucouleurs). These are not fit to the data. I cannot overstate that this is a Fourier comparison and therefore a good guess of the kind of profile. However, if you need a first idea while building a catalog quickly, this is a good choice. You will find yourself running source extractor *anyway* because many of the other packages use its segmentation map (which pixels belong to your galaxy and which don't) as an input. It remains pretty good at segmenting images in extended sources and point sources that have nothing to do with the extended source (foreground stars, background galaxies etc.).

STATMORPH—One of the many packages that build on the SE catalog and segmentation maps. It does a decent job of both computing the morphometrics (see Chapter 6) and a single Sérsic model. Mind you, this is not a fit but a computed value for both Sérsic index and effective radius.

Figure 2.11 shows a comparison between a Sérsic fit obtained on images (Kelvin et al. 2012) and those computed by STATMORPH (Pearson et al. 2019). There is broad agreement but both the Sérsic index and effective radius can be wildly off when computed by STATMORPH. The STATMORPH values are good starting points for a GALFIT fit as well as a reasonably statistical first estimate.

Figure 2.10. UGC 2885, a giant spiral galaxy about 72 Mpc distant. Some 250 kpc lt-yr across compared to the Milky Way's diameter of 30 kpc. UGC 2885 is about 10 times the size and mass as our Milky Way yet appears unruffled by recent mergers or tidal interactions. UGC 2885 was part of astronomer Vera Rubin's pioneering study of the rotation of spiral galaxies. Her work and that of A. Bosma at the radio astronomy group in Groningen showed convincing evidence of dark matter from the galaxy rotation. Image courtesy of NASA, ESA and B. Holwerda (University of Louisville) from https://www.nasa.gov/feature/goddard/2020/nasas-hubble-surveys-gigantic-galaxy.

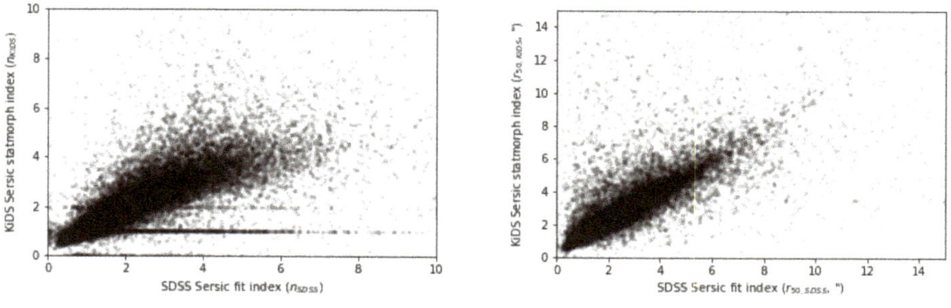

Figure 2.11. A comparison between a GALFIT Sérsic fit and those by STATMORPH. There is a wide spread in Sérsic index with a tendency for STATMORPH to over-estimate the Sérsic index a bit and there are clear alias lines in STATMORPH effective radius estimates (it defaults to an internal value). Figure from Holwerda (2022, in preparation) based on Kelvin et al. (2012) and Pearson et al. (2019) catalogs.

2.8.1 Detailed Fits

There are three detailed galaxy fit programs in general use: GALFIT was the industry standard for a long time but it can be temperamental and output depends on input starting values (see Lange et al. 2016, for a comprehensive listing of issues). It does however have a lovely set of functions to describe both bulge and disk as well as smaller features such as spiral arms, bars, central sources etc. GALFIT3 is described in Peng et al. (2002, 2010). Its benefits are speed, versatility of source descriptors, relatively easy to learn, and inter-compatibility with other results obtained with the same package. A good example of its application is the one on the S^4G survey (Salo et al. 2015) with a descriptive elements for bulge, disk and bar where applicable.

GIM2D (Simard 1998) is a slightly older package and not in as general use. However, when it is, it has been successfully deployed on a large part of the SDSS image catalog (Bottrell et al. 2019).

IMFIT (Erwin 2015) is the latest addition to the detailed package suite. Its input files are clearly influenced by GALFIT but it is more stable and allows for a variety of optimizers, including MCMC. The learning curve is a bit steeper in that it has many options that will have to be decided on.

2.9 New Sérsic Fitters

One of the developments in recent years is the availability of multi-color deep and wide imaging of very high quality. So rather than having a good image of a single or few galaxies in one or two filters, the problem is to fit a lot in a mostly automated way. In part this motivated the two packages statmorph and source extractor to calculate Sérsic values. But there are better ways: the PROFOUND/PROFIT or the GALAPAGOS/MEGAMORPH packages.

2.9.1 Profit and Profound

PROFOUND (Robotham et al. 2018) and PROFIT (Robotham et al. 2017) are entirely new approaches to galaxy-wide Sérsic profile fits. Profound is the source detection package that supplants source extractor. It detects sources in (noisy) images, generates segmentation maps identifying the pixels belonging to each source, and measure basic statistics like flux, size and ellipticity. The key difference is that source flux is measured based on the segmentation map, grown (diluted) by a function to catch all the source's flux (comparable with source extractor isomeasurements but with a correction for flux outside the segmentation map).

These segmentation maps are fed to PROFIT, the 2D profile fitter companion. Profiles can be any one of two-dimensional profile, including Sérsic, Core-Sérsic, broken-exponential, Ferrer, Moffat, empirical King, and point-source. It still works on a single source in a single filter and as such is an alternative for GALFIT.

A big advantage for detailed fits on few galaxies (or a lot but with a super-computer) is that one can use the robust (but computationally expensive) Markov Chain Monte Carlo (MCMC) approach. This allows one to identify degeneracies in the corner plots as well as accurate (and realistic) errors on the profile parameters. Figure 2.12 shows the traditional approach to Sérsic fits: data, model, data-model as a visual aid to the quality of the fit. It is already clear some spiral structure (not modeled) is left in the residual. Figure 2.13 shows the triangle plot (every fit

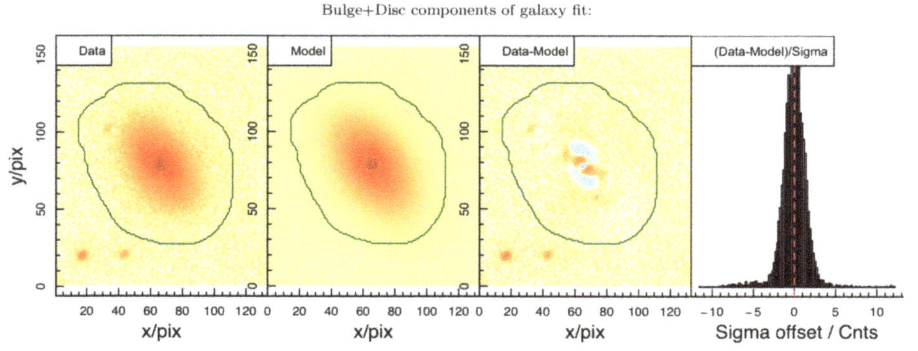

Figure 2.12. Result of the r-band MCMC decomposition of G266033 with Profit: the data (left), the bulge-disk model (second left), the data-model (second right), and the histogram of residuals (right). Even with an optimal bulge-disk model, there is still evidence of residual structure. Figure adapted from Robotham et al. (2017). Copyright © 2016 The Royal Astronomical Society.

parameter shown against the others with distributions on the diagonal) of the double Sérsic fit of a single galaxy with Profit shown in Figure 2.12. Figure 2.13 only shows stationary samplers (i.e., no "burn-in" of the sampling allowing the samplers to "walk" for better sampling). There is some evidence of weak covariance between a couple parameters (e.g., disk magnitude, Dmag, and effective radius, Dre), in general the posteriors look well sampled and consistent with normal (Gaussian) distributions. This was a well-behaved fit with little or no degeneracies. Diagnostic plots such as these allow one to assess the quality of the Sérsic 2D fit on an individual galaxy.

The integrated approach between source detection and source fits, well-documented, and open source means that these two packages can be deployed on large survey areas without much issue (other than computational expense). Examples of their use are on KiDS observations (Moffett et al. 2019) and low surface brightness galaxies (Prole et al. 2018).

2.9.2 Galapagos and MegaMorph

A fully independent approach is one which uses GALAPAGOS (Barden et al. 2012) for detection, a multi-wavelength approach using GALFIT to fit a Sérsic profile to a galaxy in multiple filters at the same time, combining the information from each filter's fit to constrain the index and effective radius. The multiband fitter is called MEGAMORPH (Häußler et al. 2013). MEGAMORPH allows only a certain change in fit values from filter to filter (e.g., a white light fit where all indices etc. are forced to be the same or a gradual polynomial change from one wavelength to the next). Figure 2.14 shows a few examples of single and MEGAMORPH fits across wavelengths. By combining information across filters, all the information is used for the final morphological fit, a distinct benefit given that most modern optical surveys include typically five filters (the SDSS *griz*) and often more.

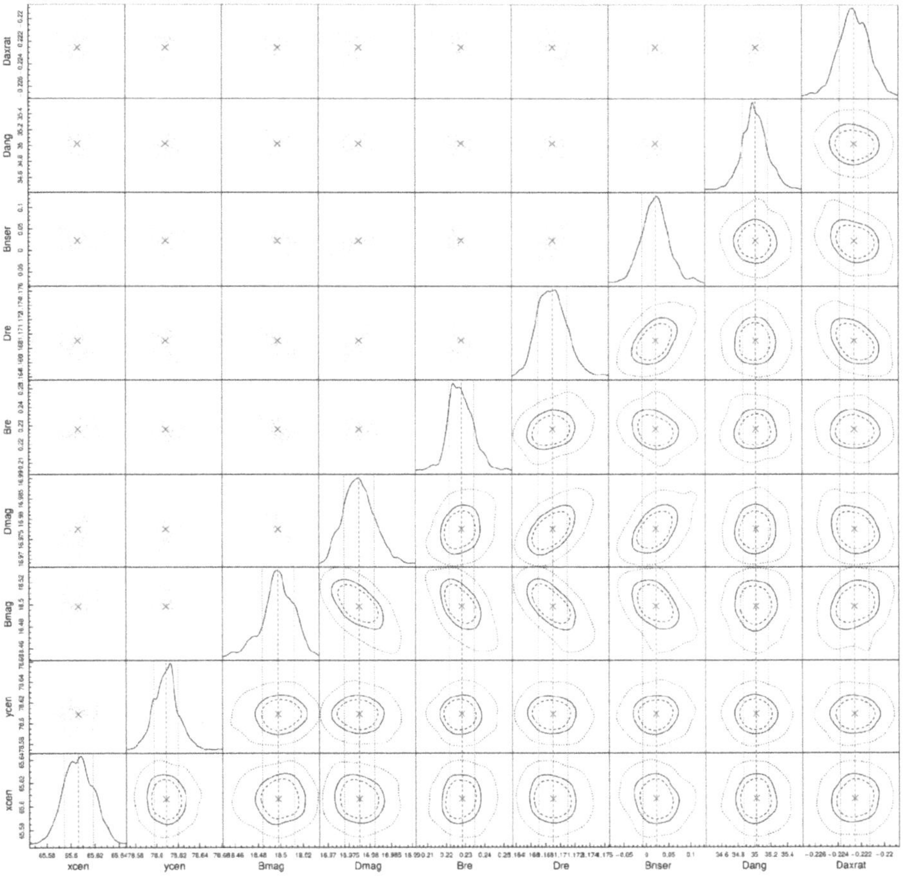

Figure 2.13. The triangle plot of the stationary MCMC chains for the model fit of G266033 shown in Figure 2.12. The top-left of the triangle shows the raw sample points, the bottom right shows the contoured version of the samples, with dashed/solid/dotted lines containing 50/68/95% of the samplers. The diagonal one-dimensional density plots show the marginalized distributions for each parameter. In this case, nine parameters were fitted using CHARM (in Profit) in the LaplacesDemon function: the position (*xcen*, *ycen*), the bulge magnitude (*Bmag*), the disk magnitude (*Dmag*), R_e of the bulge (*Bre*), R_e of the disk (*Dre*), the Sérsic index of the bulge (*Bnser*), the rotation angle θ of the disk (*Dang*), and the axial ratio of the disk (*Daxrat*). From the triangle plot, it is clear that in this case the bulge magnitude (*Bmag*) and disk magnitude (*Dmag*) show the most covariance with other parameters, but in general this fit is well converged. Note the covariances seen tend to be galaxy specific, and might not be present in generic galaxy modeling cases. Figure reprinted from Robotham et al. (2017). Copyright © 2016 The Royal Astronomical Society.

2.10 Sérsic Classifying of Galaxies

It is very tempting to think of Sérsic fits of a single way to classify galaxies. A single Sérsic fit's index immediately will tell us if a galaxy is mostly spheroid or disk right? If the index is low, it is a disk galaxy and if it is close to four it must be spheroidal.

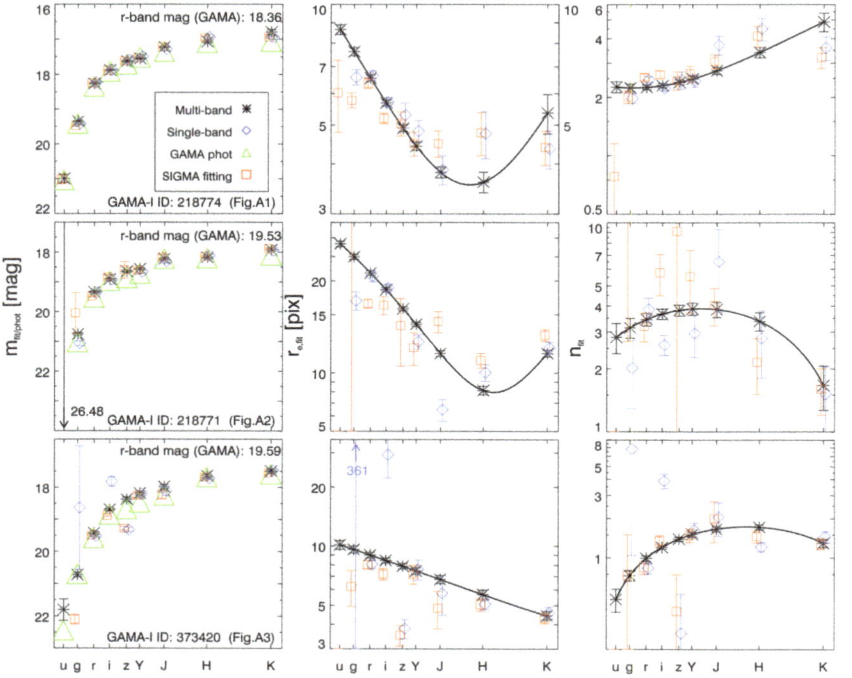

Figure 2.14. Three examples of MegaMorph multi-filter Sérsic fits to GAMA data. For each object the recovered magnitude, the effective radius and Sérsic index in each filter is shown. The black asterisks show multiband results, the blue diamonds show results from single-band fits, and orange boxes are single-band fit from Kelvin et al. (2012). The error bars are from GALFIT. The fit uses all the optical filters (*griz*) and four near-infrared filters. By constraining the change with wavelength for both fit parameters (effective radius and index) to a 3rd degree polynomial, catastrophic fit failures in noisier filters are avoided and reliability overall improves. Figure reproduced from Häußler et al. (2013). Copyright © 2013 The Royal Astronomical Society.

Even better if we have two Sérsic components by definition: a disk and a bulge. If most of the flux is fit by a disk, disk galaxy, if most of it is fit by the bulge component, it must be spheroidal (E or S0). With two components, one can classify disk galaxies by the fraction of their light that is in the bulge (spheroidal) component. This is commonly referred to as the bulge-to-total fraction (B/T).

Häußler (2021, in preparation) warns against this most strenuously. While popular in the literature, the two-Sérsic component classification approach breaks down in those galaxies that have a very weak or no second component: these are essentially disk-only or spheroid-only galaxies in the data. Because the prior for the fit routine is set to have two components, a single solution is effectively designed out: the B/T values become quite noisy.

A similar problem occurs when one tries to fit a two- or multi-component galaxy (spiral disk galaxy with prominent bulge and perhaps a bar) with a single Sérsic fit. Forcing a single profile will result in large residuals (a poor fit) but also because all the deviations are toward more flux on top of the disk the resulting Sérsic index is a noisy split between bulge and disk value.[8]

Think of it as your prior adding noise to your classification. Either you chose single Sérsic fits and those noise up anything that really has two visible components or more, or you used a double Sérsic fit which scatters single visible component galaxies throughout your *B*/*T* distribution.

2.11 Assignments

There are two jupyter notebooks to go with this chapter: Inclination Effects (Jupyter notebook 3) and Bulge Types (Jupyter notebook 4). One could do the spiral arm assignment with this chapter already (Jupyter notebook 5) or save it for Chapter 4.

2.12 Further Reading

The treatises by Graham & Driver (Graham et al. 2005; Graham & Driver 2005) on the Sérsic profile are a great reference if you want to understand those better.

References

Barden, M., Häußler, B., Peng, C. Y., McIntosh, D. H., & Guo, Y. 2012, MNRAS, 422, 449
Bottrell, C., Simard, L., Mendel, J. T., & Ellison, S. L. 2019, MNRAS, 486, 390
Breda, I., Papaderos, P., & Gomes, J.-M. 2020, A&A, 640, A20
Caon, N., Capaccioli, M., & D'Onofrio, M. 1993, MNRAS, 265, 1013
Ciotti, L. 1991, A&A, 249, 99
Courteau, S. 1996, ApJS, 103, 363
de Jong, R. S. 1996a, A&ASS, 118, 557
de Jong, R. S. 1996b, A&A, 313, 45
de Jong, R. S. 1996c, A&A, 313, 377
de Jong, R. S., & van der Kruit, P. C. 1994, A&ASS, 106, 451
de Vaucouleurs, G. 1948, AnAp, 11, 247
de Vaucouleurs, G. 1959, AJ, 64, 397
Erwin, P. 2015, ApJ, 799, 226
Graham, A. W., & Driver, S. P. 2005, PASA, 22, 118
Graham, A. W., Driver, S. P., Petrosian, V., et al. 2005, AJ, 130, 1535
Grudić, M. Y., Hopkins, P. F., Quataert, E., & Murray, N. 2019, MNRAS, 483, 5548
Häußler, B., Bamford, S. P., Vika, M., et al. 2013, MNRAS, 430, 330
Hopkins, P. F., Croton, D., Bundy, K., et al. 2010, ApJ, 724, 915
Jarrett, T. H., Koribalski, B. S., Kraan-Korteweg, R. C., et al. 2007, AJ, 133, 979
Kado-Fong, E., Petrescu, M., Mohammad, M., et al. 2021, ApJ, 920, 72
Kassin, S. A., de Jong, R. S., & Pogge, R. W. 2006a, ApJS, 162, 80
Kassin, S. A., de Jong, R. S., & Weiner, B. J. 2006b, ApJ, 643, 804
Kelvin, L. S., Driver, S. P., Robotham, A. S. G., et al. 2012, MNRAS, 421, 1007
Lange, R., Moffett, A. J., Driver, S. P., et al. 2016, MNRAS, 462, 1470
Ling, C.-T., Hashimoto, T., Goto, T., et al. 2020, MNRAS, 496, 864
McGaugh, S. S., & Bothun, G. D. 1994, AJ, 107, 530

[8] Think Jean-Claude van Damme kind of split. Impossible and likely painful.

McGaugh, S. S., Bothun, G. D., & Schombert, J. M. 1995, AJ, 110, 573
Moffett, A. J., Phillipps, S., Robotham, A. S. G., et al. 2019, MNRAS, 489, 2830
Pastrav, B. A., Popescu, C. C., Tuffs, R. J., & Sansom, A. E. 2013, A&A, 553, A80
Pearson, W. J., Wang, L., Alpaslan, M., et al. 2019, A&A, 631, A51
Peng, C. Y., Ho, L. C., Impey, C. D., & Rix, H.-W. 2002, AJ, 124, 266
Peng, C. Y., Ho, L. C., Impey, C. D., & Rix, H.-W. 2010, AJ, 139, 2097
Pohlen, M., Zaroubi, S., Peletier, R. F., & Dettmar, R.-J. 2007, MNRAS, 378, 594
Prole, D. J., Davies, J. I., Keenan, O. C., & Davis, L. J. M. 2018, MNRAS, 478, 667
Robotham, A. S. G., Davies, L. J. M., Driver, S. P., et al. 2018, MNRAS, 476, 3137
Robotham, A. S. G., Taranu, D. S., Tobar, R., Moffett, A., & Driver, S. P. 2017, MNRAS, 466, 1513
Salo, H., Laurikainen, E., Laine, J., et al. 2015, ApJS, 219, 4
Sérsic, J. L. 1963, BAAA, 6, 41
Sérsic, J. L. 1968, Atlas de Galaxias Australes (Cordoba: Observatorio Astronomico)
Simard, L. 1998, in ASP Conf. Ser. 145, Astronomical Data Analysis Software and Systems VII ed. R. Albrecht, R. N. Hook, & H. A. Bushouse (San Francisco, CA: ASP), 108
van der Kruit, P. C., & Searle, L. 1981, A&A, 95, 105
Vulcani, B., Bamford, S. P., Häußler, B., et al. 2014, MNRAS, 441, 1340

Chapter 3

Galaxy Sizes

In the galaxy-wide characterization of light distribution of the extended objects that are galaxies, we already encountered the effective radius used in the Sérsic profile (Chapter 2). This is not the only scale used for galaxy size and we will discuss the different ones here. Keep in mind that different scales were defined with different uses in mind: photometry (reliably catching a representative amount of light), or a relation to a physical scale (e.g., the virial radius). Table 3.1 summarizes the most commonly used ones.

3.1 Photometric Radii

Photometric radii have been the longest in use so it is important to know about them from a historical perspective if for no other reason. Different photometric radii were defined to ensure a reliable fraction of the light would be included for photometry.

3.1.1 R_{25}

The R_{25} (and D_{25}) is the radius (diameter) of the aperture defined by 25 mag arcsec^{-2} in the B-band filter. This scale is most easily defined as one is inspecting galaxies on photographic plates. This value is reported in the Third Reference Catalogue of Bright Galaxies (RC3) by de Vaucouleurs et al. (1991). This is a scale that is guaranteed to be available for nearby galaxies and sometimes therefore in use to put all galaxies in a sample on the same footing.[1]

The RC3 attempts to be reasonably complete for galaxies having apparent diameters larger than 1′ at this D_{25} isophotal level and total B-band magnitudes brighter than 15.5, and a redshift not in excess of 15,000 km s^{-1}.

[1] I did in my thesis but at the time (2000–2005) it already felt a bit…quaint.

doi:10.1088/2514-3433/ac2c7dch3

Table 3.1. Radii Nomenclature

Radius	Symbol	Description
Kron	R_k	The typical size of the flexible aperture—computed from moments—defined by Kron (1980).
Petrosain	R_p	The radius at which the surface brightness of the isophote is η times the average surface brightness within this isophote.
Effective	R_e	The radius containing 50% of the total flux of an object.
Half-light		Same as the above radius.
Typical	R_t or h	The scale in the exponential disk $I = I_0 \exp(-R/R_t)$
de Vaucouleurs	R_{25}	The radius at which the B-band isophote is 25 mag arcsec^{-2}. This is a common radius as it is listed in the third reference catalog (RC3)
Mass threshold	R_1	The radius where galaxies reach $M_\odot/pc^2$.

3.1.2 Holmberg Radius

Very similar as the D25, the Holmberg radius is defined as 26 mag arcsec^{-2} in sdss-g. It is an isophotal radius, often abbreviated as R_H.

3.1.3 Petrosian Radius

The Petrosian radius is another photometry radius (a scale linked to counting the light from a galaxy). In this case, Petrosian (1976) defined a point in the radial light profile at which the isophote at that radius was a certain fraction of the *average* surface brightness within that radius.

$$\eta(R) = \frac{2\pi \int_0^R I(R')R'\,\mathrm{d}R'}{\pi R^2 I(R)} = \frac{\langle I \rangle_R}{I(R)} \tag{3.1}$$

where the parameter η is the fraction. This is often either $\eta = 0.2$ or $\eta = 0.5$ with $\eta = 0.2$ the most commonly used. However, for example, $\eta = 0.5$ is used in source extractor (SE).

Depending on the fraction of the surface brightness (η) chosen, a few times the Petrosian radius will encapsulate most of the flux of a galaxy (see Figure 3.1). From Figure 3.1 it is immediately evident why at a lower value of η helps to catch a more representative fraction of the flux of a galaxy. But especially for elliptical/spheroidal ($n > 2$) profiles, the light (and therefore the number of stars in that galaxy) is underestimated by some 10%–20%! This rapidly adds up if one wants to estimate the total mass in stars in for example a cluster of galaxies (a bunch of elliptical galaxies together).

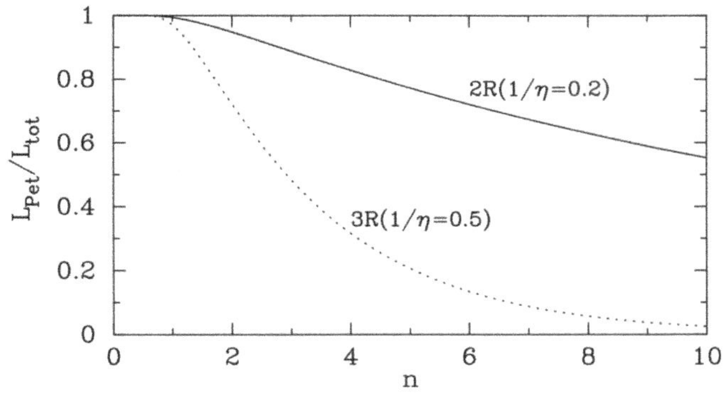

Figure 3.1. The fraction of the total flux (L_{tot}) captured by the Petrosian radius as a function of light-profile shape, the Sérsic n. The Petrosian luminosity L_{Pet} within the Petrosian radius for $\eta = 0.2$ and $2 \times R_p(\eta = 0.2)$ (solid line) and for $\eta = 0.5$ and $3 \times R_p(\eta = 0.5)$ (dotted line). Figure reproduced with permission from Graham & Driver (2005).

3.1.4 Kron Radius

Kron (1980) defined the Kron[2] radius to get 90% of an objects light as follows:

$$R_l(R) = \frac{2\pi \int_0^R I(x)x^2 \, dx}{2\pi \int_0^R I(x)x \, dx}. \tag{3.2}$$

This is an estimate based on the second order moment of the light; flux weighted by the position squared (x^2). The most common use of this radius is in the source extractor (SE). According to Graham et al. (2005), the original SE publication has changed the definition a bit. SE (Bertin & Arnouts 1996) define the Kron radius as:

$$R_l = \frac{\Sigma R \ I(R)}{\Sigma I(R)} \tag{3.3}$$

over the two-dimensional aperture and not for a light profile (which was Kron's original idea). In the case of a light-profile (already radially averaged), the Kron radius should be according to Kron's original definition:

$$R_l = \frac{\Sigma R^2 \ I(R)}{\Sigma R \ I(R)}. \tag{3.4}$$

The Kron radius coming out of SE is maybe not a proper one but SE is using this as the estimate anyway. Note how integrations become summations when dealing with pixel collections.

And if you are interested in galaxy profiles, here is a neat trick to relate this Kron radius to the *effective* radius (R_e), the radius that encloses half the object's light.

[2] Named after him. At least it's not a law.

$$R_l(x, n) = \frac{R_e}{b^n} \frac{\gamma(3n, x)}{\gamma(2n, x)}. \tag{3.5}$$

In which $x = b(R/R_e)^{1/n}$ and γ is the incomplete gamma function.[3] This is all too much of a headache so Graham et al. (2005) also give a nice table converting these radii depending on Sérsic profile index (copied in Table 3.2). In the case of a disk galaxy (Sérsic n=1), $2 \times R_l$ catches 90% of the light (Figure 3.1). Scaling up to $2.5 \times R_l$, the Kron radius catches consistently 90+% of the flux of a galaxy, see Figure 3.2.

Table 3.2. Theoretical Kron Radii and Magnitudes from Graham et al. (2005)

Sérsic n	R_l (R_e)	$L(<2R_l)$%	$L(<2.5R_l)$%
0.5	1.06	95.7	99.3
1.0	1.19	90.8	96.0
2.0	1.48	87.5	92.2
3.0	1.84	86.9	90.8
4.0	2.29	87.0	90.4
5.0	2.84	87.5	90.5
6.0	3.53	88.1	90.7
7.0	4.38	88.7	91.0
8.0	5.44	89.3	91.4
9.0	6.76	90.0	91.9
10.0	8.39	90.6	92.3

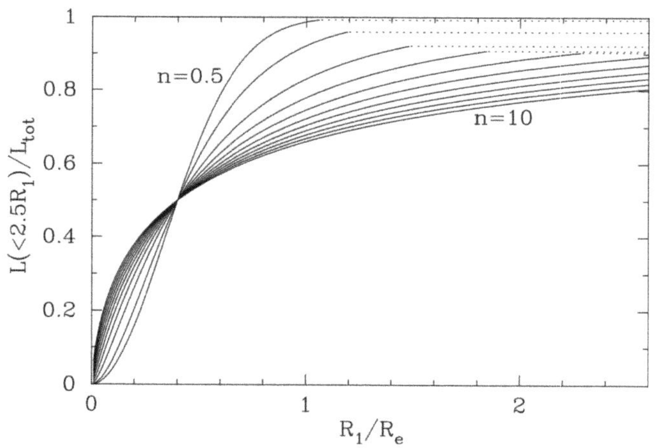

Figure 3.2. The Kron radius performance (aperture set to $2.5 \times R_l$): fraction of the total flux as a function of effective radius (R_e) and for each Sérsic profile (n). Figure reproduced with permission from Graham & Driver (2005).

[3] Oh Gamma function, is there something you're not useful for?

3.2 Effective Radius

The half-light radius or effective radius is now the most common way to describe the size of a galaxy. It remains fairly robust with different image depths and is an integral part of the Sérsic radius description.

As we have seen in the previous section, the Kron radius and Petrosian radii were developed for photometry. The effective radius is a profile variable and it is often used in the galaxy luminosity or mass with size relations, see Figure 3.3.

> **Half-light or Effective Radius?** The half-light and "effective" radius are often used inter-changeably and…mostly they are. If the profile is a pure Sérsic then half the light is enclosed at the effective radius ($r_{50} = r_{\rm hl} = R_e$). However, if more than Sérsic component is involved, R_e refers to half the light in that component! The galaxy's half-light refers more to the aperture in which half the light is enclosed using curve-of-growth photometry! Half-light radii contain half of all the light of an object. Effective radius just for a given morphological component. So in the case of a multi-component galaxy, the two are not the same.

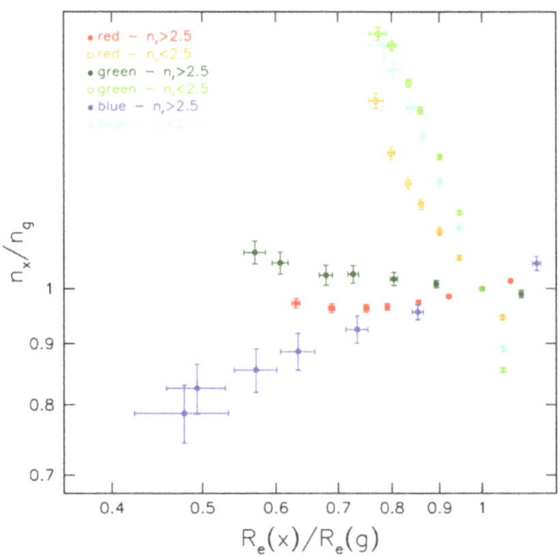

Figure 3.3. Summary plot of multi-wavelength (i.e., in many different filters) single Sérsic profile fits (Vulcani et al. 2014). Each point is the average of Sérsic values for a filter, normalized on the values in the sdss-g filter. Colors denote the broad colors of the galaxies (red/blue/green) and if they are considered disk galaxies ($n_{\rm r}$ < 2.5) or spheroidal galaxies ($n_{\rm r}$ > 2.5). This is the same information shown in Figures 2.8 and 2.9, now combined into just $R_{\rm e}$ versus n. Spheroidal galaxies have no big change in Sérsic index especially for red and green populations but the greatest change is in effective radius and disk galaxies do have a significant change in Sérsic profile and only a small change with wavelength in effective radius. Figure reproduced from Vulcani et al. (2014). Copyright © 2014 The Royal Astronomical Society.

3.3 Isomass Radius (R1)

Rather than an isophotal or a profile-dependent size, one could select a stellar density threshold to define the size of a galaxy with. There is a physical motivation behind this: we think there has to be a threshold surface density in a galaxy to allow for star formation.[4] If one starts with that, one quickly arrives at $\sim 1 M_{\odot}/pc^2$. this makes for an easy-to-remember and reasonable density threshold to define a galaxy size on. Figure 3.4 shows the issue quire clearly: the effective radii depends strongly on what type of galaxy is considered. The R1 value much better delineates the size of each object. Taking this further, Figure 3.5 shows the mass–size relation using the effective radius and the R1. The correlation is much tighter with R1 (less scatter) and shows a steady increase. This looks to be a good indication of galaxy size.

There are some caveats though. As noted, one needs a conversion from the light profile into a mass surface density profile reliably. For optical colors, one can take a fiducial stellar population and use a color term to convert a surface brightness into a stellar mass density (Bell & de Jong 2001; Zibetti et al. 2009, 2010). The effects of

Figure 3.4. Examples of galaxies showing the location of their R_e and R1, the isomass contour at $1 M_{\odot}/pc^2$. The top row shows elliptical galaxies (E0-S0+), the middle row shows spiral galaxies (S0/a-Sm), and the bottom row shows dwarf galaxies. Galaxy mass increases from left to right. The isomass radius (R1) nicely encloses the bulk of the stellar mass of every type of galaxy. The colored regions are the g, r, and i band composite, while the black and white background is the sum of the three bands. The background level of these images is $\sim$29.1 mag arcsec^{-2}. Figure reproduced from Trujillo et al. (2020). Copyright © 2020 The Royal Astronomical Society.

[4] This is sometimes called the "Kennicutt–Schmidt Law" (Kennicutt 1998). Apparently we weren't quite done with laws in astronomy in the late 1990s.

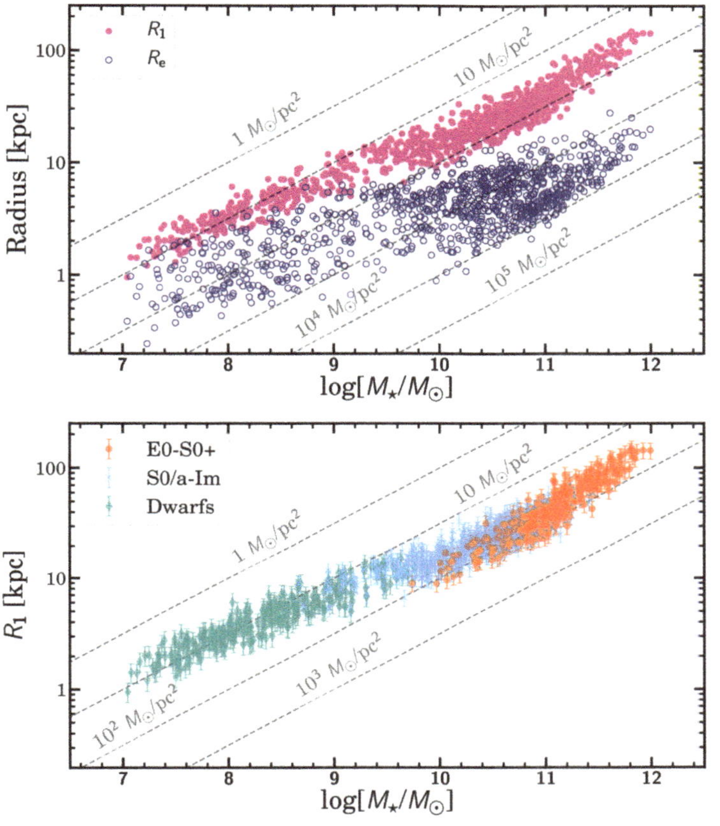

Figure 3.5. Stellar mass–size relations: the observed R1–mass and R_e–mass relations (upper panel). The scatter of the relation using R1 is significantly smaller compared to that with R_e (sdss-g). The same R1–mass relation by category: spheroidals/ellipticals (E0-S0+), spirals (S0/a-Im), and dwarf galaxies. The gray dashed lines correspond to locations in the plane with constant (projected) stellar mass density. Disk galaxies follow a different relation (slope) than spheroids (ellipticals). Figure reproduced from Trujillo et al. (2020). Copyright © 2020 The Royal Astronomical Society.

stellar population are the greatest in bluer filters (e.g., sdss-g) so the issue with effective radius becomes smaller if one measures in redder or infrared filters.

But now that we are moving to ever lower surface brightness regimes, uncovering many more dwarf galaxies, R1 is undoubtedly more suitable as a measure of galaxy size.

Higher Redshift Measuring the sizes of galaxies at higher redshifts rapidly brings the issue of rest-frame or emitted wavelength. The issue with using effective radius may be fine for nearby galaxies using red or near-infrared colors but a redshifted galaxy will be visible in that wavelength only in the mid or far-infrared! So size comparisons in blue filters (and even ultraviolet filters) will be done and necessary and these will need a comparison local equivalent. So we measure sizes in every filter, see Figure 3.6.

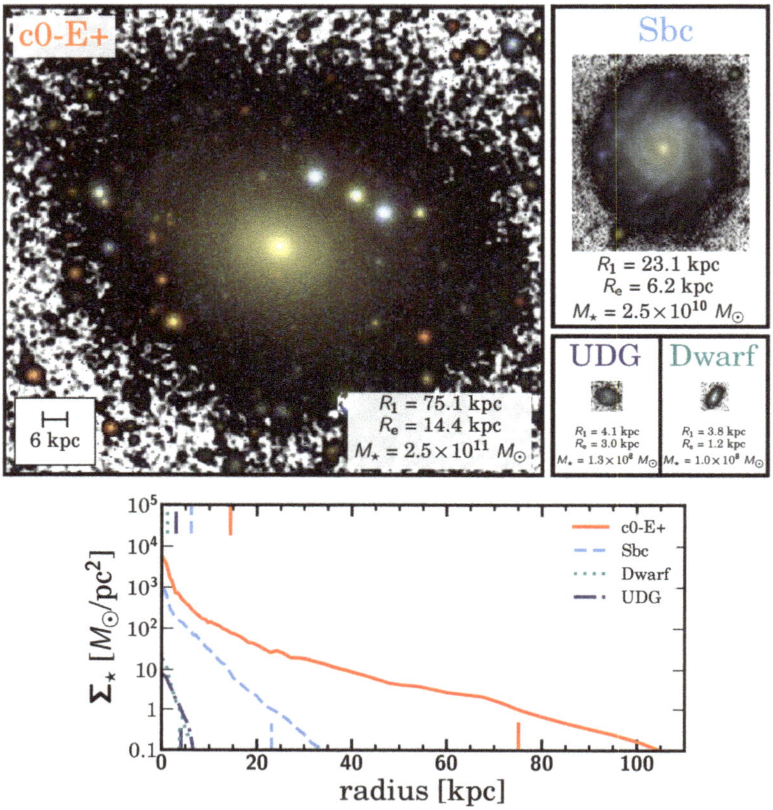

Figure 3.6. Top: representative galaxies of different stellar masses: a giant elliptical ($M^{*}{\sim}2.5 \times 10^{11} M_{\odot}$), a Milky Way-like galaxy ($M^{*}{\sim}2.5 \times 10^{10} M_{\odot}$), a ultra diffuse galaxy (UDG) and dwarf galaxy ($M^{*}{\sim}10^{8} M_{\odot}$). All galaxies are shown to the same physical scale and a similar depth in surface brightness. Bottom: stellar mass density profiles of the same galaxies. The colored ticks in the top x-axis mark the location of Re. Similar ticks in the lower x-axis mark the location of R1. Figure reproduced with permission from Chamba et al. (2020) © ESO.

3.4 Relating Galaxy Scales

How are the scale length, R_{25} or $R_{\rm H}$, the Kron and Petrosian radii and half-light radius related to each other?

Keep in mind that any relation implies an assumed radial light profile: it is ok to average out in ellipses how the light declines from the center of the galaxy. In effect this ignores galaxy substructures (arms, bars etc. see Chapter 4) and lopsidedness (Section 4.7). Most of the time, it assumes only a single Sérsic profile (i.e., we ignore the fact that disk galaxies have a bulge in the center).

This has been explored for some of the scales, especially the relation between effective radius and Petrosian radius (Graham et al. 2005; Graham & Driver 2005), see Figure 3.7.

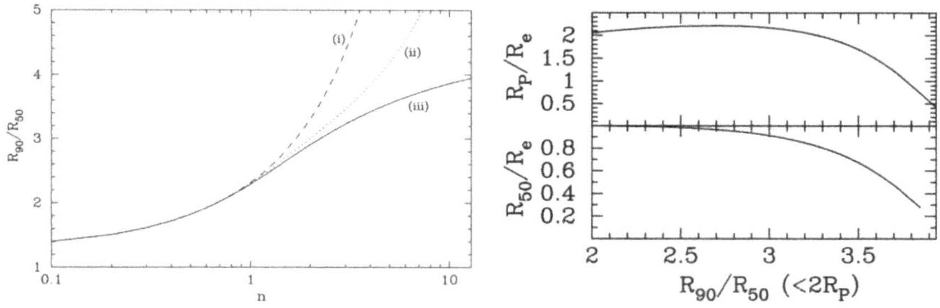

Figure 3.7. Left: the relation between the radii where 90% (outer) and 50% of the light enclosed by 2x the Petrosian radius are reached for different Sérsic profile indices (n). The steeper the profile, the higher this concentration index is (higher n, more concentrated light profile, see Concentration definitions in Chapter 6). Right: the ratio of the Petrosian radius over the Sérsic profile's effective radius (R_P/R_e) and the 50% radius (half-light radius i.e., the radius of the light enclosed in the Petrosian radius) over the effective radius of the Sérsic profile. Note that these are not united and that the ratio depends on the profile. Figure reproduced from Graham et al. (2005).

3.4.1 Size–Luminosity

A single Sérsic profile can be fit to galaxy images (see earlier) and this has been done extensively to the Sloan Digital Sky Survey and every digital available survey since. Figure 3.8 shows the luminosity–size relation for Sloan galaxies, expressed as the Sérsic effective radius and absolute magnitude. Disk galaxies have a shallower (almost) power law relation between luminosity and effective radius. The spheroidal (elliptical and S0) galaxies do have a power law (straight line in log–log space in Figure 3.8) and much less spread in that relation.

3.4.2 Size–Mass

Luminosity is tied to total stellar mass but not in a straightforward manned (it depends on the stellar population and the Initial Mass Function assumed). Sloan magnitudes can be converted to mass estimates, using a color (cf Zibetti et al. 2004; Bell & de Jong 2001; or Taylor et al. 2011). Once one has a mass estimate, one can now map the mass–size relation. More massive galaxies are bigger is not a shocking statement but it is notable that spheroidal galaxies still behave as a power law relation and the disk galaxies show a more subtle behavior. Figure 3.9 shows the stellar mass and size relation for Sloan galaxies (Shen et al. 2003).

3.4.3 Size Evolution; The Early Growth of Galaxies

The Sloan results shown before are true for the local universe but how did galaxies grow? To do that, we measure galaxy sizes, often via Sérsic profile fits, at the same wavelength as we did in Sloan. This means going to the near-infrared with space-based imaging for high redshift. Near-infrared deep observations with the Hubble and Spitzer Space Telescopes as well as ground-based surveys have resulted in a boon in the numbers of high-redshift galaxies ($z > 6$) identified by the Lyman break in their optical and near-infrared colors.

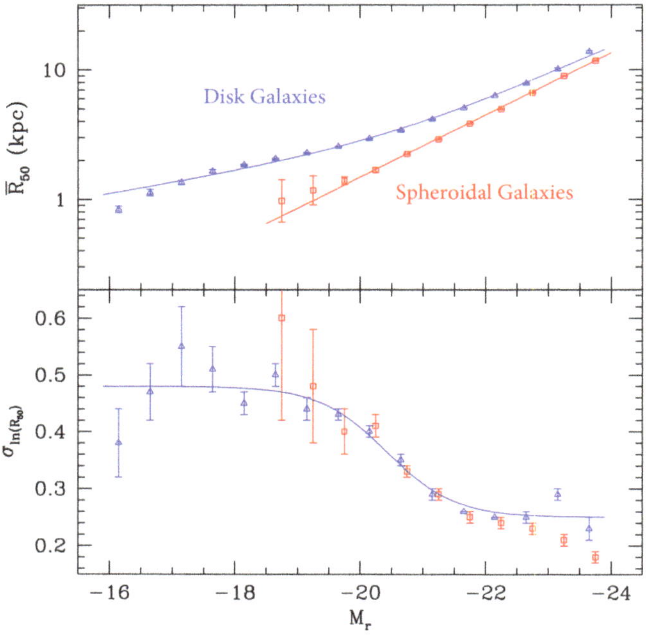

Figure 3.8. The luminosity–size relation for Sloan Digital Sky Survey galaxies. Luminosities and sizes measured in the sdss-r band (red). The galaxies are divided in disk-dominated galaxies (concentration $c > 2.86$) and spheroidal dominated galaxies (concentration $c < 2.86$). The mean effective radius is shown in the top panel and the variance in the population shown in the bottom panel. Brighter galaxies are to the right of the figure. Disk galaxies's size is not quite a power law with luminosity and shows much greater variance at lower luminosities. Spheroidal (elliptical) galaxies show a clear power law but it does not extend all the way to where the disk galaxies do. Figure adapted from Shen et al. (2003). Copyright © 2003 The Royal Astronomical Society.

The high-redshift frontier is now firmly at $z \sim 9$–10, the limit of Hubble with the Lyman dropout technique, with a dozen high-confidence galaxies known (Zheng et al. 2012; Coe et al. 2013; Bouwens et al. 2011b, 2011a, 2013; Ellis et al. 2013; Oesch et al. 2013, 2014). These highest redshift galaxies can be identified by their extremely red near-infrared colors ($J - H > 0.5$), a lack of flux in bluer (optical) bands, and relatively blue H-4.5 μm colors.[5] The fainter $z \sim 9$–10 candidates were found both behind lensing clusters (Coe et al. 2013; Zheng et al. 2012), and in ultra-deep WFC3/IR observations (Bouwens et al. 2011b; Ellis et al. 2013; Oesch et al. 2013).

The brightest objects at these redshifts are exceedingly rare but a number of them have been identified in both the Hubble deep fields (Oesch et al. 2014) and more recently in shallower Hubble observations (Calvi et al. 2016; Morishita et al. 2018). The existence of such bright galaxies at these redshifts pose a critical challenge for

[5] The Spitzer Space Telescope observed for a long time in two wavebands: 3.5 and 4.5 micron. These never got a letter to denote them.

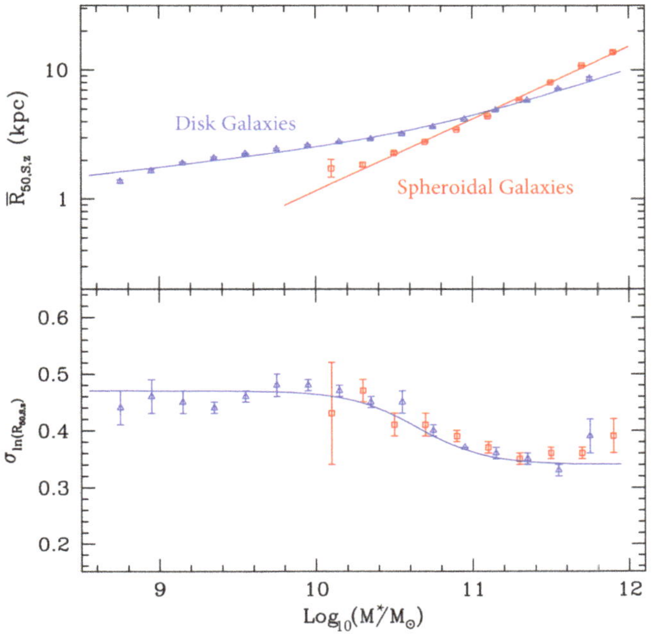

Figure 3.9. The size–stellar mass relation for disk and spheroidal galaxies. Same distinction as in Figure 3.8. Disk galaxies are closer to a power law (straight line) but not quite. Figure adapted from Shen et al. (2003). Copyright © 2003 The Royal Astronomical Society.

early galaxy evolution because there is only some 500 Million years to build it up (Madau & Dickinson 2014; Bouwens et al. 2015; Ishigaki et al. 2018). One check if these objects are indeed high-redshift galaxies is to check their sizes against what one expects. Comparing these sizes against model expectations for luminous galaxy candidates at $z \sim 9$–10 and the sizes of potential interlopers (Grazian et al. 2012; Holwerda et al. 2015).

The analytical models (Fall & Efstathiou 1980; Mo et al. 1998) predict effective radii should scale with redshift somewhere between $R_e \propto (1+z)^{-1}$ for galaxies living in halos of fixed mass or $R_e \propto (1+z)^{-1.5}$ at a fixed circular velocity. Observational evidence from earlier samples also points to such scaling relations, with some studies preferring $R_e \propto (1+z)^{-1}$ (Bouwens et al. 2004, 2006; Oesch et al. 2010), some studies preferring $R_e \propto (1+z)^{-1.5}$ (Ferguson et al. 2004), and some studies lying somewhere in between (Hathi et al. 2008; Ono et al. 2013; Shibuya et al. 2015). For the bright ($>0.3L^*$) sources, one expects galaxies to follow $R_e \propto (1+z)^{-1.5}$ relation (Holwerda et al. 2015), see Figure 3.10.

3.5 Assignments

There are two Jupyter notebook assignments to go with galaxy sizes: Galaxy Size Measurements (Jupyter notebook 6) and Sizes of Galaxies (Jupyter notebook 7).

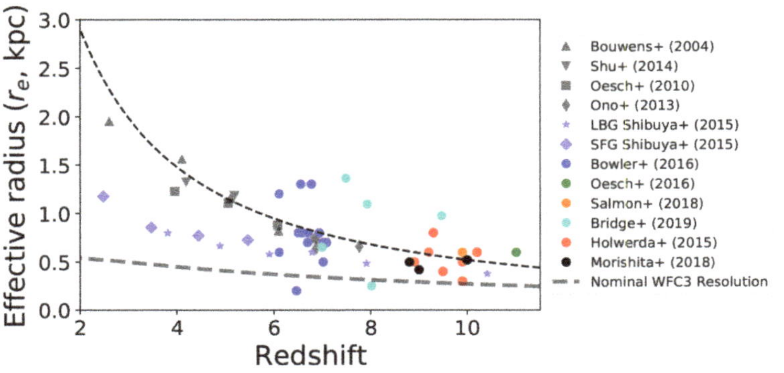

Figure 3.10. Early galaxy size evolution: effective radius as a function of redshift for the bright galaxies from several previous studies: *mean* effective radii (Bouwens et al. 2004; Zheng et al. 2014; Oesch et al. 2010; Ono et al. 2013; gray symbols). The mean values of the size distributions for the star-forming galaxies (SFG) and Lyman break galaxies (LBG) for comparison (light blue symbols; Shibuya et al. 2015). The effective radii as a function of redshift for *individual* galaxies from Holwerda et al. (2015; $z \sim 9$), Bowler et al. (2017) and Bridge et al. (2019; $z \sim 8$), and the $z \sim 9$–10 galaxies identified by Morishita et al. (2018) and Holwerda et al. (2020; circles). The dashed line is the best fit to $r_e \propto (1 = z)^m$ with $m = -1.32$ (Holwerda et al. 2015), a value somewhere in between the two evolutionary scenarios ($m = -1$ or $m = -1.5$). The mean values are for bright galaxies while the mode is measured over the whole observed luminosity range but is dominated by lower luminosity ($L < 0.3L^*$) galaxies. The individual objects at the higher redshifts are predominantly bright and show a wide spread around the mean for bright sources. The gray dashed line is the Hubble infrared camera, the Wide Field Camera 3's nominal resolution. Figure reproduced from Holwerda et al. (2020).

3.6 Further Reading

The relationships between galaxy size measurements is treated exhaustively in Graham & Driver (2005). Higher redshift sizes are comprehensively done in Shibuya et al. (2015) and Oesch et al. (2007) is a good primer for high-z size measurements.

References

Bell, E. F., & de Jong, R. S. 2001, ApJ, 550, 212
Bertin, E., & Arnouts, S. 1996, A&ASS, 117, 393
Bouwens, R. J., Illingworth, G. D., Blakeslee, J. P., Broadhurst, T. J., & Franx, M. 2004, ApJL, 611, L1
Bouwens, R. J., Illingworth, G. D., Blakeslee, J. P., & Franx, M. 2006, ApJ, 653, 53
Bouwens, R. J., Illingworth, G. D., Labbe, I., et al. 2011a, Natur, 469, 504
Bouwens, R. J., Illingworth, G. D., Oesch, P. A., et al. 2011b, ApJ, 737, 90
Bouwens, R. J., Illingworth, G. D., Oesch, P. A., et al. 2015, ApJ, 803, 34
Bouwens, R. J., Oesch, P. A., Illingworth, G. D., et al. 2013, ApJL, 765, L16
Bowler, R. A. A., Dunlop, J. S., McLure, R. J., & McLeod, D. J. 2017, MNRAS, 466, 3612
Bridge, J. S., Holwerda, B. W., Stefanon, M., et al. 2019, ApJ, 882, 42
Calvi, V., Trenti, M., Stiavelli, M., et al. 2016, ApJ, 817, 120
Chamba, N., Trujillo, I., & Knapen, J. H. 2020, A&A, 633, L3

Coe, D., Zitrin, A., Carrasco, M., et al. 2013, ApJ, 762, 32
de Vaucouleurs, G., de Vaucouleurs, A., Corwin, H. G., et al. 1991, Third Reference Catalogue of Bright Galaxies, Vol. 1–3 (Berlin: Springer-Verlag)
Ellis, R. S., McLure, R. J., Dunlop, J. S., et al. 2013, ApJL, 763, L7
Fall, S. M., & Efstathiou, G. 1980, MNRAS, 193, 189
Ferguson, H. C., Dickinson, M., Giavalisco, M., et al. 2004, ApJL, 600, L107
Graham, A. W., & Driver, S. P. 2005, PASA, 22, 118
Graham, A. W., Driver, S. P., Petrosian, V., et al. 2005, AJ, 130, 1535
Grazian, A., Castellano, M., Fontana, A., et al. 2012, A&A, 547, A51
Hathi, N. P., Jansen, R. A., Windhorst, R. A., et al. 2008, AJ, 135, 156
Holwerda, B. W., Baldry, I. K., Alpaslan, M., et al. 2015, MNRAS, 449, 4277
Holwerda, B. W., Bridge, J. S., Steele, R. L., et al. 2020, AJ, 160, 154
Ishigaki, M., Kawamata, R., Ouchi, M., et al. 2018, ApJ, 854, 73
Kennicutt, R. C. Jr 1998, ApJ, 498, 541
Kron, R. G. 1980, ApJS, 43, 305
Madau, P., & Dickinson, M. 2014, ARA&A, 52, 415
Mo, H. J., Mao, S., & White, S. D. M. 1998, MNRAS, 295, 319
Morishita, T., Trenti, M., Stiavelli, M., et al. 2018, ApJ, 867, 150
Oesch, P. A., Bouwens, R. J., Illingworth, G. D., et al. 2010, ApJL, 709, L16
Oesch, P. A., Bouwens, R. J., Illingworth, G. D., et al. 2013, ApJ, 773, 75
Oesch, P. A., Bouwens, R. J., Illingworth, G. D., et al. 2014, ApJ, 786, 108
Oesch, P. A., Stiavelli, M., Carollo, C. M., et al. 2007, ApJ, 671, 1212
Ono, Y., Ouchi, M., Curtis-Lake, E., et al. 2013, ApJ, 777, 155
Petrosian, V. 1976, ApJL, 209, L1
Shen, S., Mo, H. J., White, S. D. M., et al. 2003, MNRAS, 343, 978
Shibuya, T., Ouchi, M., & Harikane, Y. 2015, ApJS, 219, 15
Taylor, E. N., Hopkins, A. M., Baldry, I. K., et al. 2011, MNRAS, 418, 1587
Trujillo, I., Chamba, N., & Knapen, J. H. 2020, MNRAS, 493, 87
Vulcani, B., Bamford, S. P., Haussler, B., et al. 2014, MNRAS, 441, 1340
Zheng, W., Postman, M., Zitrin, A., et al. 2012, Natur, 489, 406
Zheng, W., Shu, X., Moustakas, J., et al. 2014, ApJ, 795, 93
Zibetti, S., Charlot, S., Rix, H.-W., et al. 2009, MNRAS, 400, 1181
Zibetti, S., Charlot, S., & Rix, H.-W. 2010, in IAU Symp. 262, Stellar Populations—Planning for the Next Decade (Cambridge: Cambridge Univ. Press), 89
Zibetti, S., White, S. D. M., & Brinkmann, J. 2004, MNRAS, 347, 556

Chapter 4

Sub-galaxy Scale Elements

This chapter is for the features that define individual galaxies but are not on the scale of the galaxy as a whole (10's of kpc) but typically bigger than star clusters (<kpc).

4.1 Bulges

We start with bulges but the central spheroidal component to a disk galaxy is its second most distinctive feature (apart from the disk). Initially it was thought that these central spheroids all followed the de Vaucouleurs ($n = 4$ Sérsic) profiles and could be viewed akin to a small elliptical in the center of a larger disk galaxy. These, we now call "classical bulges" and parts of the disk that were heated kinematically into a central component, often red or yellow in color because of the older (and metal poorer) stellar component but following a light profile that is closer to a disk ($n = 1 - 2$ Sérsic profiles). We call this kind "pseudo-bulges." The upshot is that not every bulge is equal and some may be flatter than others as well as some are "pressure supported" (randomized orientations of all the stellar orbits, like in an elliptical) and some are still rotating (and therefore flatter as well).

4.1.1 Classical Bulges

In the classical view (Hubble Sequence!) there is the disk and a secondary component, the bulge. In this traditional view, the bulge is treated as something similar to a small elliptical galaxy embedded within the larger disk: more spheroidal than the disk, with stellar motions dominated by velocity dispersion rather than rotation, and having a strongly concentrated structure (a $1/R^4$ surface brightness profile like that of an elliptical). An example of a classical bulge is shown in Figure 4.1.

In addition, the stellar populations of bulges seem to be like those of ellipticals: (older, more metal-rich, and alpha-enhanced) than the majority of stars in the disk. Taken all together, this seemed to argue for a bulge formation mechanism similar to

doi:10.1088/2514-3433/ac2c7dch4

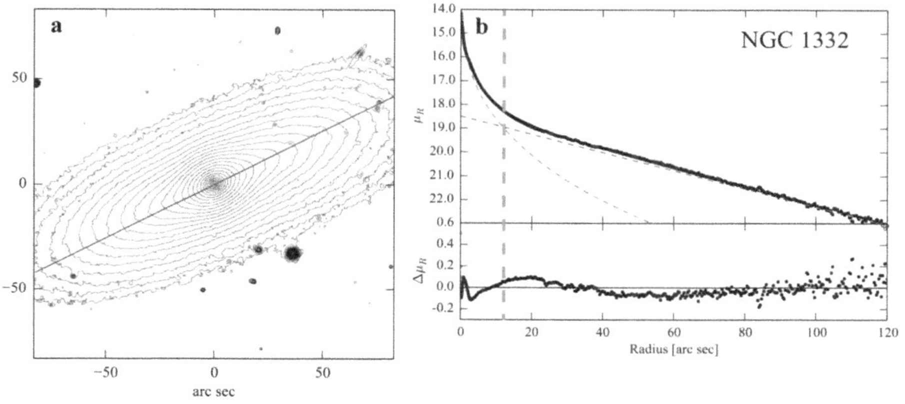

Figure 4.1. The classical bulge in the low-mass S0 galaxy NGC 7457. (a) Log-scaled V-band isophotes (JKT image); gray line marks major axis (PA = 126°). (b) Bulge-disk decomposition of major-axis profile. Data (black circles) combine major-axis cuts through HST WFPC2 F555W image ($r < 6.6$ arcsec) and ground-based V-band image. Dashed lines are Sérsic + exponential fit to the data, with residuals in lower sub-panel. Vertical dashed green line marks bulge/disk radius $R_{\rm bd}$, where Sérsic and exponential components are equally bright, the boundary of the "photometric bulge." Figure reproduced from Erwin et al. (2015). Copyright © 2014 The Royal Astronomical Society.

that proposed for ellipticals, either via monolithic collapse of a giant proto-galaxy cloud or by rapid, violent mergers of initial sub-components at high redshift.

4.1.2 Pseudo-bulges

Pseudo-bulges are also central components of disk galaxies but the main conjecture is that these have formed from the disks themselves through internal processes and not through violent mergers. An example of a pseudo-bulge is shown in Figure 4.2.

If secular processes built dense central components that masquerade as bulges, how can one distinguish pseudo-bulges from merger-built bulges? Pseudo-bulges have one or more characteristics of disks: (a) flatter shapes than those of classical bulges (an exponential rather than $1/R^4$), (b) correspondingly large ratios of ordered to random velocities (they still rotate), (c) smaller stellar velocity dispersions σ than one would expect for a bulge of that size (see the Faber–Jackson relation), (d) spiral structure or nuclear bars in the "bulge" part of the light profile, and (f) evidence for ongoing or recent star formation (Kormendy & Kennicutt 2004).

4.1.3 Observed Bulges

With the Sloan Digital Sky Survey observations, it was possible to fit a bulge + disk decompositions of millions of galaxies (Gadotti 2009; Simard et al. 2011). Figure 4.3 shows the distribution of Sérsic indices for bulges. There is a clear range of values, spanning "classical" and "pseudo-bulges." Seen in detail, there is evidence that classical bulges are not simply an elliptical galaxy surrounded by a disk but an intermediate form between ellipticals of the same mass and pseudo-bulges (Gadotti 2009), see Figure 4.4.

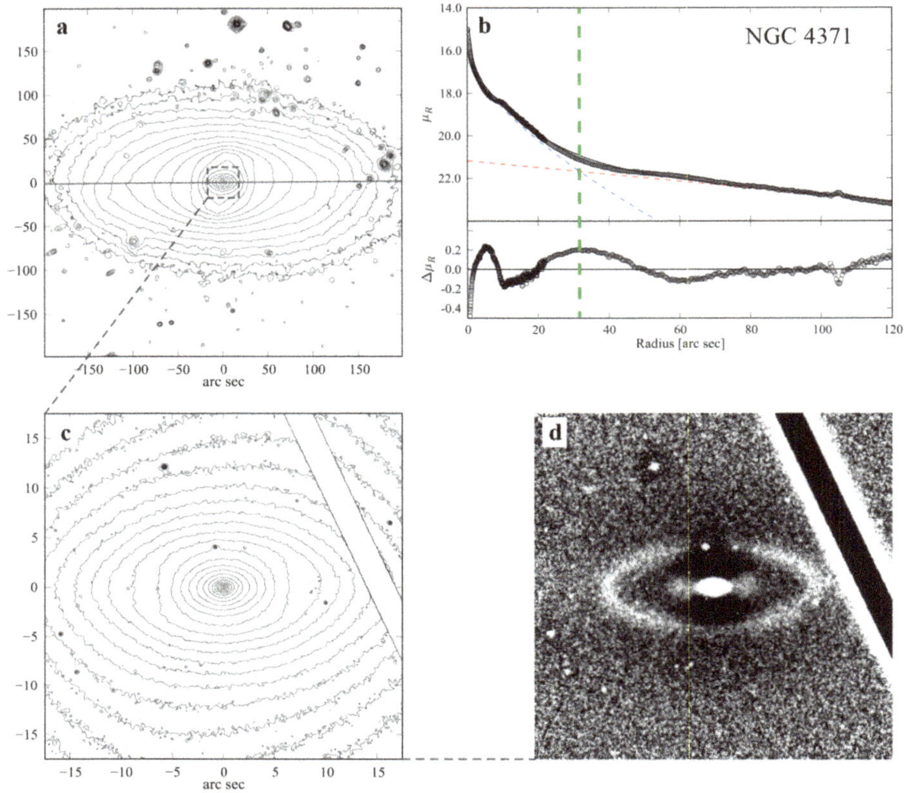

Figure 4.2. The pseudo-bulge in the S0 galaxy NGC 4371. (a) Log-scaled R-band isophotes (Isaac Newton Telescope); gray line marks major axis (PA = 90.4°). (b) Bulge-disk decomposition of INT major-axis profile. Dashed lines are a Sérsic + exponential fit to the data, with residuals plotted in lower sub-panel. Vertical dashed green line marks "bulge/disk" radius R_{bd}. Figure reproduced from Erwin et al. (2015). Copyright © 2014 The Royal Astronomical Society.

4.2 Bulge Formation

Currently, the two leading possible bulge-formation scenarios have been proposed, each to explain a type of bulge: one for classical bulges and one for pseudo-bulges.

First, the "early" bulge formation scenario predicts that mergers of small systems in the early universe result in the formation of a spheroidal, pressure supported (i.e., random orbits of the stars) system. Following this early formation, a disk grows around the bulge through gas accretion and subsequent star formation. A bulge formed in this manner is dynamically hot (random stellar motions), smooth and featureless, and very similar to a major merger remnant, i.e., an Elliptical galaxy (Fisher & Drory 2008).

Second, the "late-bulge formation" scenario proposes that disks form first and then bulges form from them through *in-situ* events in the disk, such as disk instabilities (e.g., a bar) and epicyclic motions. The disk instabilities lead to the flow of gas toward the center of the gravitational potential and epicyclic motions

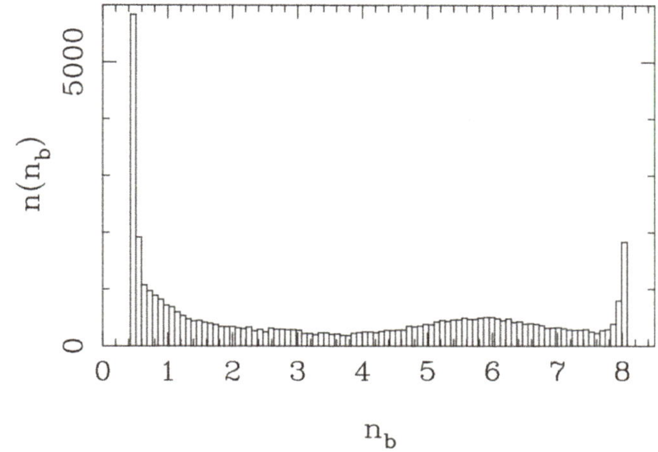

Figure 4.3. Distribution of GIM2D bulge Sérsic index n_b from the free n_b bulge + disk decompositions for galaxies with bulge to total $B/T > 0$. Figure reproduced from Simard et al. (2011). © 2011. The American Astronomical Society. All rights reserved.

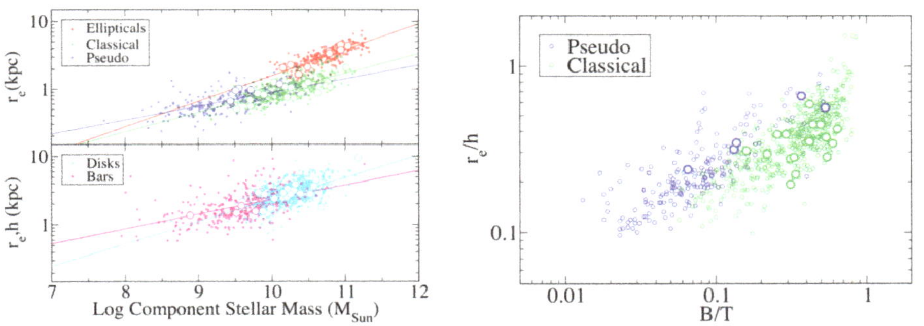

Figure 4.4. Left: the scale lengths of elliptical galaxies, classical bulges, pseudo-bulges, disks and bars as a function of the corresponding component's stellar mass. White-filled circles represent systems with $0.5 < b/a < 0.7$ (face-on), with the same color coding. Right: the ratio between the scale lengths of bulges and disks re/h against bulge-to-total B/T for pseudo-bulges and classical bulges. Pseudo-bulges and classical bulges follow offset relations. For the same B/T, pseudo-bulges have on average higher values of re/h. Figure reproduced from Gadotti (2009). © 2009 The Author. Journal compilation © 2009 RAS.

amplify over time once the disk is stable, causing *centralized* star formation and the growth of a bulge inside the already established disk. This type of bulge is dynamically cold (still mostly rotating; Kormendy 1993; Gao et al. 2020). In terms of stellar population (color and metallicity), these bulges are much more similar to their disk than the classical bulges (Fisher 2006; Du et al. 2020; Gao et al. 2020).

Based on these scenarios, it was initially thought that classical and pseudo-bulges could be distinct in their Sérsic index: pseudo-bulges would have Sérsic indices close to unity ($n < 2$), i.e., a near-exponential surface brightness profile, resembling more the disk they originated from. Classical bulges would have a higher Sérsic index, $n > 2$, more akin to that of spheroid galaxies they resemble in color, appearance and origin (Andredakis & Sanders 1994; Andredakis et al. 1995; Fisher 2006; Méndez-Abreu et al. 2010; Mendez-Abreu 2016).

However, recent kinematic decomposition studies find that Sérsic index is not a good representative of different types of bulge (Krajnović et al. 2013; Zhu et al. 2018; Schulze et al. 2018). So the proposed neat morphological distinction may not be so neat after all!

4.3 Spiral Arms

Astronomers frequently refer to spiral arms in a galaxy, but what is a spiral arm exactly? It is not a matter of semantics or "You'll know it when you see it."[1] Part of the reason is probably that defining what constitutes a visual arm is not easy (cf. Figure 4.5). Humans also find it difficult to agree on how many arms are in a spiral galaxy or what part of the galaxy is arm and inter-arm, see Figures 4.6, 4.7 and 4.8.

However visually there is definitely a more or less clear spiral pattern in most if not all galaxies. Spiral arms are classified "grand design" when they seem an unbroken swirl of stars and "flocculant" when the spiral pattern in short and broken in many sections (Elmegreen et al. 2011). UGC 2885 (Figure 2.10) or M101 (Figure 4.18) are examples of grand-design (see Figures 4.5 and 4.6 on how consensus can be found on the numbers of spiral arms).

4.3.1 Spiral Structure of the Milky Way

For forest-for-the-trees reasons, astronomy struggles to get an accurate count and map of the spiral structure of our own Milky Way. Star counts are incomplete and other tracers (e.g., molecular clouds) are difficult to get a distance to. Figure 4.9

Figure 4.5. Example images of galaxies at $z = 0.03$ and $M_r \sim -21$ with both tightly wound and loose spiral arms (upper and lower rows, respectively) and small or large bulges (left and right columns, respectively). In each case, galaxies are shown with either strong bars ($p_{\rm bar} > 0.5$) or no bar ($p_{\rm bar} < 0.2$) to the left or right in each section. Images are gri composites from SDSS with a scale of 1.7 arcmin square. Figure reproduced from Masters et al. (2019). © 2019 The Author(s), Published by Oxford University Press on behalf of the Royal Astronomical Society.

[1] One of the finer definitions given by the legal minds of the US Supreme court.

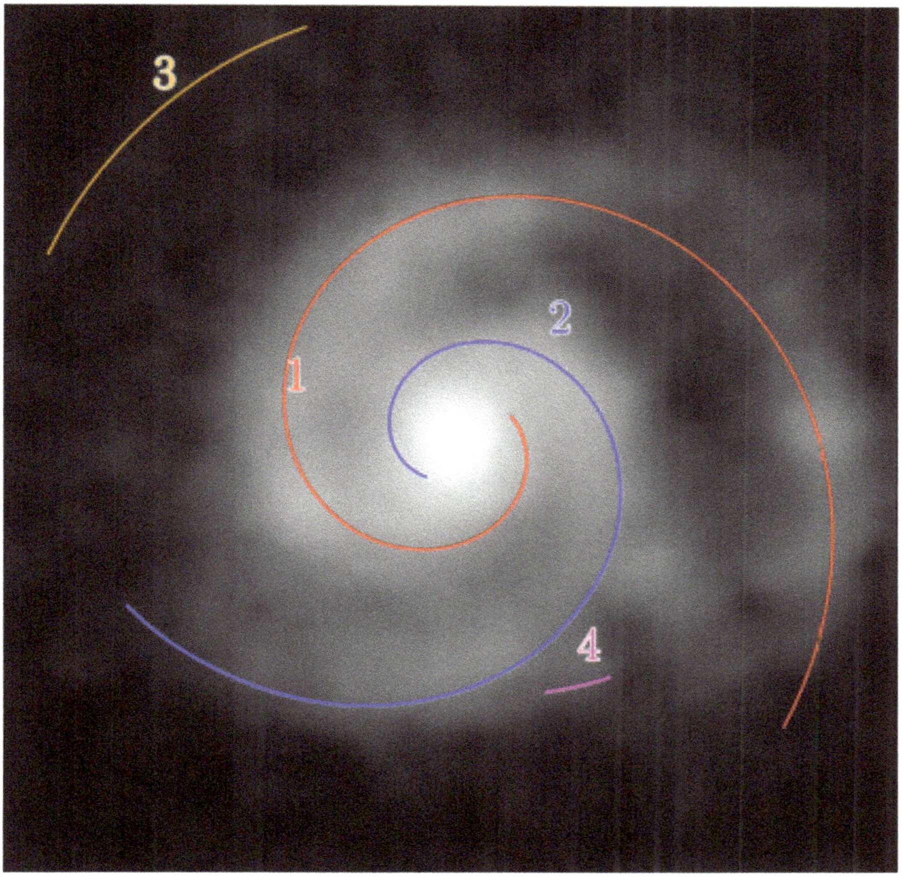

Figure 4.6. An example of a galaxy image presented to volunteers in the *spiralspotter* Galaxy Zoo project. The grayscale image is the Sloan Digital Sky Survey r-band image of the galaxy deprojected to face-on. The colored curves indicate where SPARCFIRE-identified arcs in the image, each of which are assigned a number. Each arc was assigned a unique color and number for volunteer classification. Figure reproduced from Hart et al. (2017). © 2017 The Authors, Published by Oxford University Press on behalf of the Royal Astronomical Society.

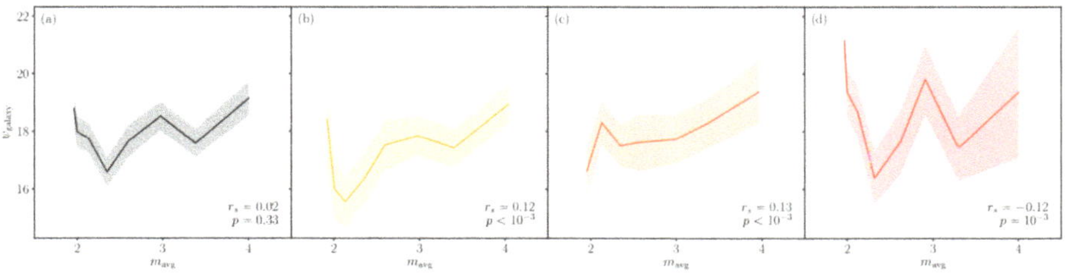

Figure 4.7. Weighted average arm number (m_{avg}) versus pitch angle (ϕ) for the stellar mass-matched spiral sample with reliable arcs for (a) all, (b) unbarred, (c) weakly barred, and (d) strongly barred galaxies. The data are divided into eight bins in each panel, and the line shows the mean for each bin. The shaded error region shows one standard error on the mean. The arms of fewer-armed galaxies are tighter for both the unbarred and weakly barred subsamples. Figure reproduced from Hart et al. (2017). © 2017 The Authors, Published by Oxford University Press on behalf of the Royal Astronomical Society.

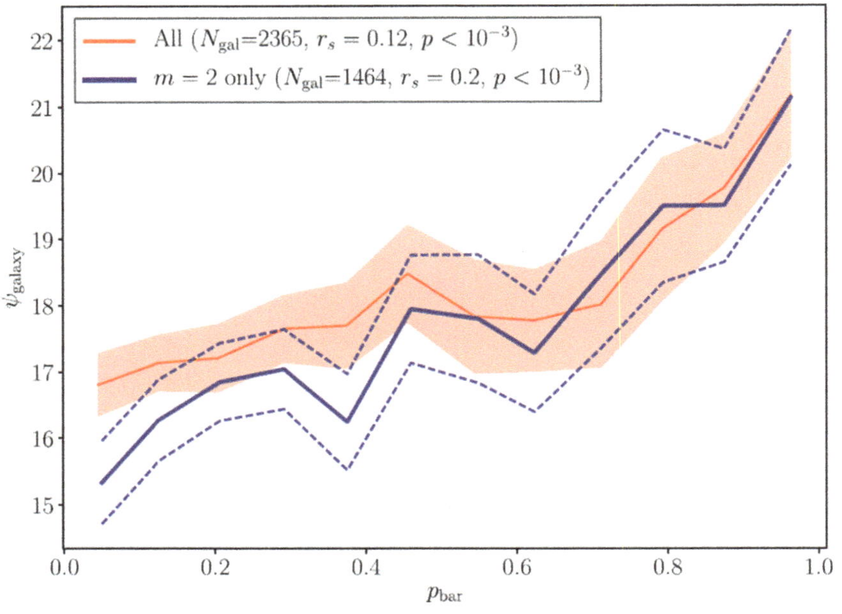

Figure 4.8. Galaxy Zoo bar fraction, pbar, versus spiral arm pitch angle (ϕ_{galaxy}). The orange line width filled errors indicates the relation for all spiral galaxies in the stellar mass-limited sample. The thicker blue line with dashed errors shows the same relation for only galaxies with two spiral arms in Galaxy Zoo ($m = 2$). The lines indicate the mean for each of the bins, and the errors show one standard error on the mean. Galaxies with stronger bars have looser spiral arms. Figure reproduced from Hart et al. (2017). © 2017 The Authors, Published by Oxford University Press on behalf of the Royal Astronomical Society.

shows the consensus at the moment: four arms, more or less connected, emanating from the bar. Figure 4.10 illustrates what the issue is for using line emission to reconstruct spiral structure: rotational velocities are super-imposed along each line of sight from the solar system. Depending on which molecule one uses, a different view of the spiral structure emerges. If one uses star counts, whole spiral arms go missing (Churchwell et al. 2009). Figure 4.11 shows that even with precise distances from the GAIA satellite and therefore a good three-dimensional map of star clusters, it remains tricky to identify how many spiral arms there are. For now we adopt the model in Figure 4.9 but this could be subject to change!

4.3.2 Spiral Structure

Spiral structure can be quantified using a logarithmic spiral to describe the shape of arms in spiral galaxies. The stellar orbits are essentially circular and that the spiral arms are the result of an evolving pattern (Ringermacher & Mead 2009). Stars congregate in a locus point at a given point of their orbits.

The simplest curve that describes galaxies is the logarithmic spiral and has been used by many (Lin & Shu 1964; Kennicutt 1981; Kennicutt & Hodge 1982; Elmegreen & Elmegreen 1987; Seigar & James 1998a, 1998b; Seigar et al. 2006, 2014) in their morphological descriptions:

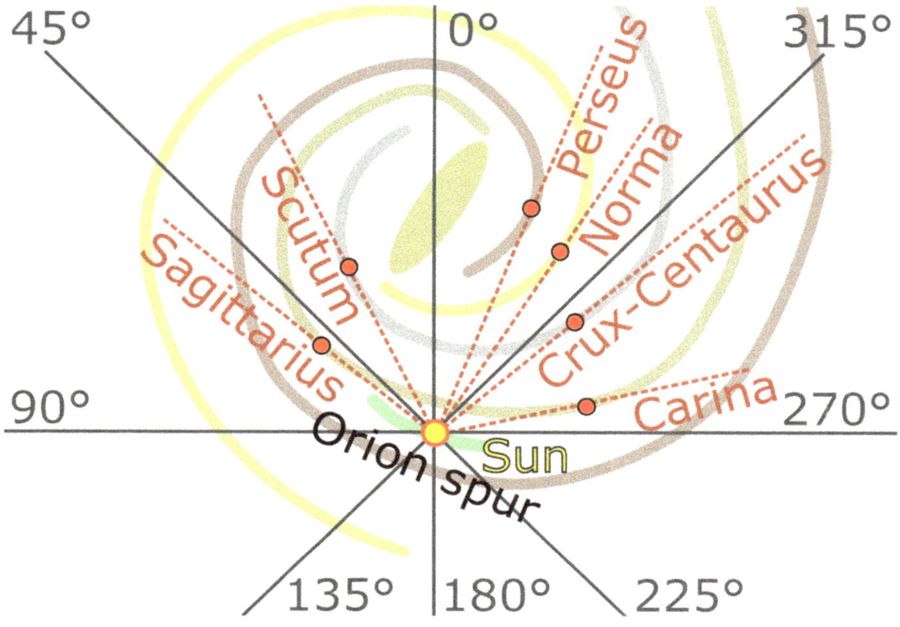

Figure 4.9. A sketch of the Milky Way's spiral structure. The spiral arms are in separate colors. The solar system is placed in the Orion spur about 8.2 kpc away from the Galactic center. Together with the Milky Way's center the Sun defines the Galactic coordinate system where angles are measured counter clockwise along the plane of the disk as longitude and angle away from the plane as latitude. Figure reproduced with permission from Reissl et al. (2020) © ESO.

$$r(\phi) = r_0 e^{k\phi} \tag{4.1}$$

where k is the constant angle of pitch and ϕ describes the angle in the disk, and r and r_0 are the radius and starting radius of the spiral arm pattern (see Figure 4.12). Note here that this is for (mostly) grand design patterns. Flocculant spiral arms could be described as well like this but only for a short interval (include a radius for which the arm ends as well).

The pitch, P, is defined from Binney & Tremaine (1987):

$$\cot(P) = r(\phi)\frac{d\phi}{\mathrm{d}r}. \tag{4.2}$$

For a first attempt, $P = \tan^{-1} k$ is constant. However, it is often apparent that galaxy arms do not have constant pitch. Just look at the arms in M101 (Figure 7.19). This is not merely describing what we observe but the spiral arm structure could well be linked to the dark matter profile in disks (Seigar et al. 2006, 2014).

The focus now has moved from detailed decomposition of spiral structure of a few nearby galaxies (Gadotti 2009; Seigar et al. 2014) to large-scale statistics of the level of winding and pitch angles (Masters et al. 2019; Walmsley et al. 2022; Figure 4.5). With more HI kinematics becoming available, the idea of linking spiral arm structure

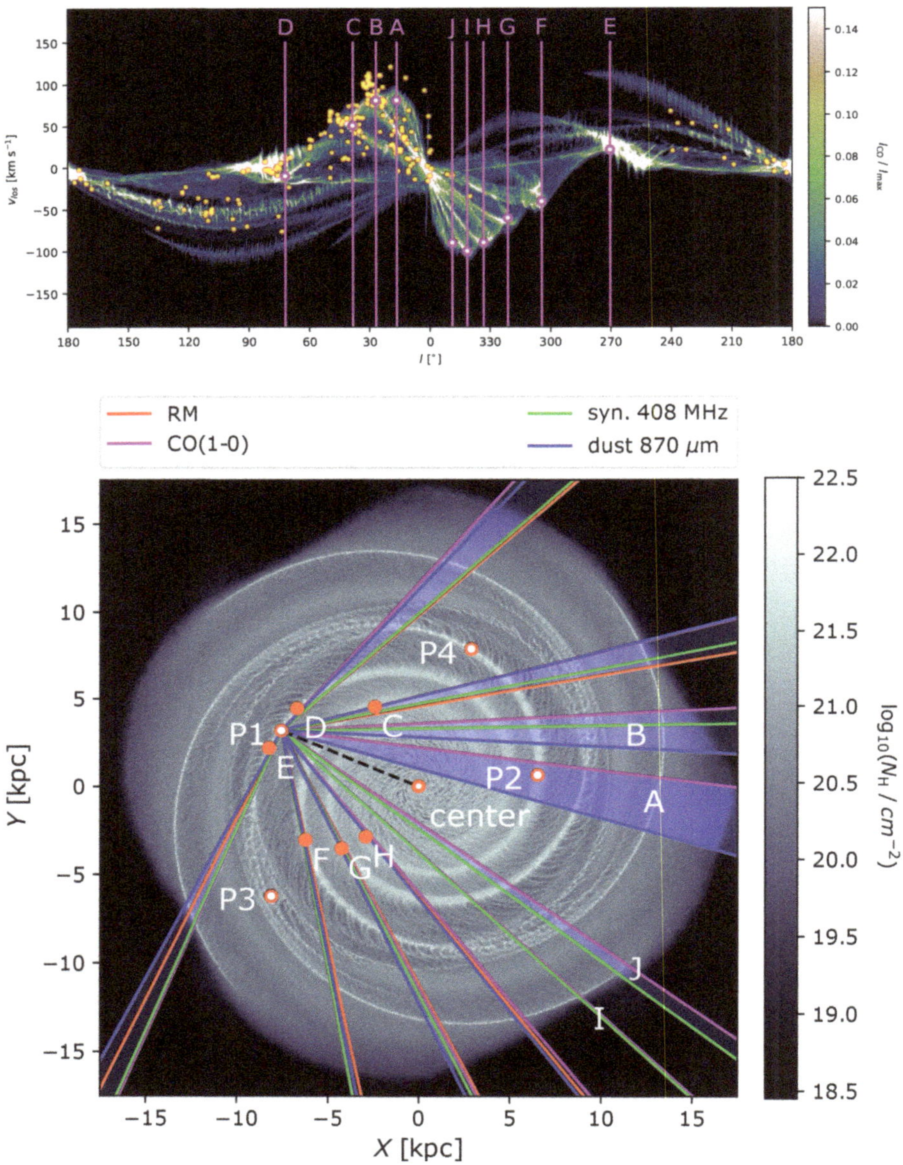

Figure 4.10. Longitude–velocity diagram of the 12CO(1-0) emission line constructed using our Galactic model by placing the observer at position P1. The velocities are averaged. Purple dots and vertical lines show detections of spiral arm tangent points as determined from CO. Yellow dots represent observations of maser emissions coming from high-mass star-forming regions within our own Milky Way presented in Reid et al. (2019). Note that detections in general depend on the tracers used. The detections are labeled A–J with increasing Galactic longitude. The magnitude of the CO intensity is normalized. Figure reproduced with permission from Reissl et al. (2020) © ESO.

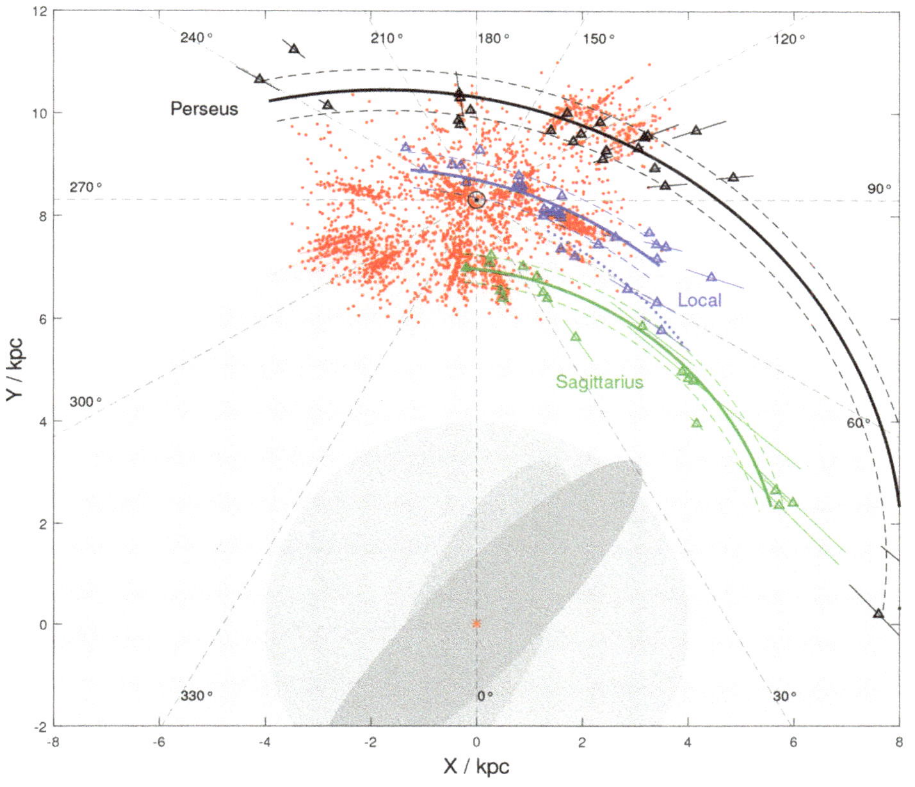

Figure 4.11. The OB (O-B2 type) stars (red circles) and masers (triangles) from Reid et al. (2009) and Xu et al. (2019). GAIA parallax uncertainties of the OB stars shown here are better than 10%. The Milky Way center (red star) is at (0,0) and the Sun (Sun symbol) is at (0, 8.31). The solid and dashed curved lines denote the center and widths of spiral arm models based on VBLI maser parallaxes. The Perseus arm (black), the Local arm (blue), and Sagittarius figure reproduced with permission from Xu et al. (2018) © ESO.

characteristics and dark matter profiles (shear and concentration) can now be explored on an industrial scale.

4.4 Bars

Bars are the most important internal perturbations seen in disk galaxies. A bar is an elongated mass often made of old stars crossing the center in highly elliptical orbits. If spiral structure is present, the arms begin at either end of the bar. Although most easily recognized in the face-on view, bars can also be detected in the edge-on view as a central thick, boxy shape.

Bars are understood as a natural instability in a rotationally supported stellar disk (Sellwood & Wilkinson 1993). The longevity of a bar depends on how well its angular momentum is transferred to other galactic components (mostly the dark matter halo; Athanassoula 2003; Collier et al. 2018).

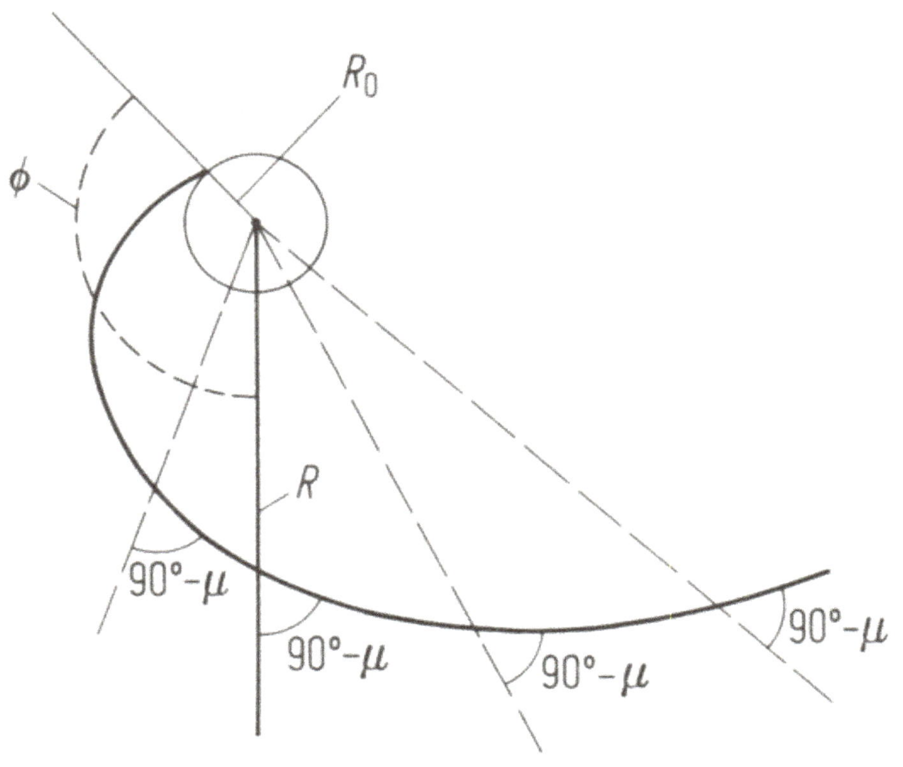

Figure 4.12. The logarithmic spiral. ϕ = winding angle; μ = pitch angle; $90° - \mu$ = characteristic angle. From https://ned.ipac.caltech.edu/level5/STRUCTURE/spst.html.

Bars may be transient features that may dissolve and regenerate several times over the age of the universe (Bournaud & Combes 2002; Collier et al. 2018). Or bars may be long-lived density wave modes that drive secular evolution of both the stellar and gaseous distributions (Zhang & Buta 2007; Buta et al. 2009; Salo et al. 2010; Knapen 2010; Díaz-García et al. 2016).

Not all bars are made exclusively of old stars: some barred galaxies have considerable (molecular) gas and star formation along the bar. Bars are considered a way to funnel gas to the center of the galaxy in order to refuel the central AGN.[2] The dust lanes associated with bars are not only the point of star formation in bars but trace inflowing gas toward the central source.

It was controversial for a while how long bars did survive in disk galaxies. Are these transient phenomena? What is the fraction of disk galaxies that have a bar? Does that change with redshift? The consensus now seems to be that massive disks and prominent bars have been around since $z \sim 1$ and most of the evolution (more barred galaxies) has happened in the lower-mass disk galaxies (Sheth et al. 2008). But over time, more disk galaxies seem to acquire a bar (Figure 4.13).

[2] Active galactic nucleus, the central supermassive black hole.

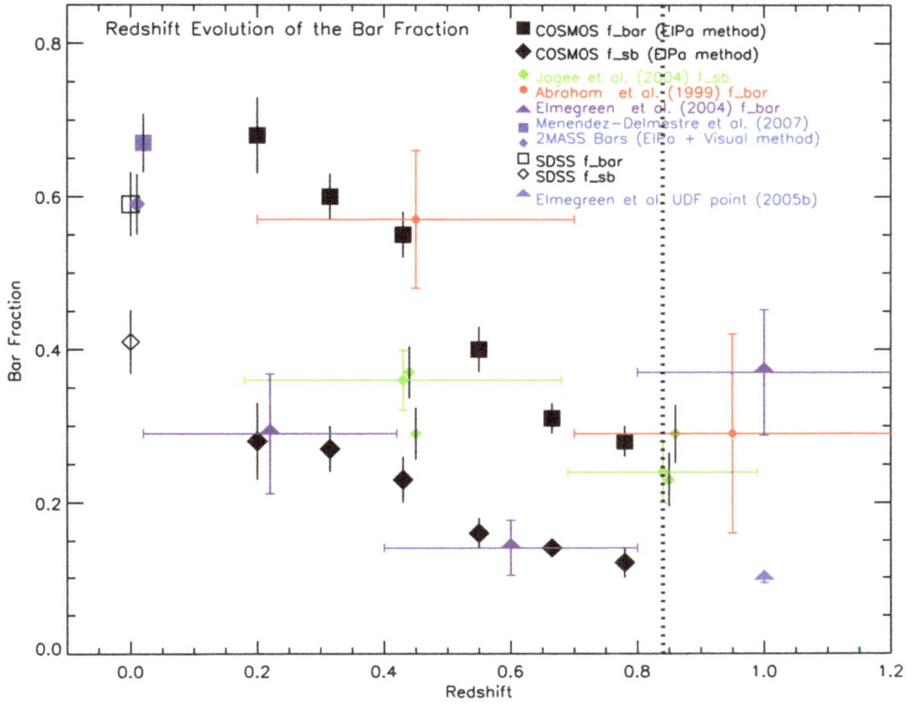

Figure 4.13. Bar fractions as a function of redshift from a series of searches over Hubble Space Telescope deep images (Abraham et al. 1999; Elmegreen et al. 2004; Jogee et al. 2004; Sheth et al. 2008). There is a notable increase with bar fraction (number of disk galaxies with a credible bar) with decreasing redshift. Figure reproduced from Sheth et al. (2008).

4.5 Lenses

Similar to bars are elongated features known as ovals or "lenses,"[3] something between a linear bar and the disk itself with a preferred axis. These ovals can be major perturbations in a galactic disk. Up to the entire disk of a galaxy may be considered an oval. Ovals are most easily detected if there is considerable starlight at larger radii to contrast the oval against.

Lenses (or ovals) are features, made usually of old stars, that have a shallow brightness gradient interior to a sharp edge. They are commonly seen in S0 galaxies. They are subtle features easily missed by automated classification. Lenses are round or slightly elliptical in shape. If elliptical in shape they could also be considered ovals. Figure 5.4 shows examples of lenses in red, blue, and green galaxies.

The origin of the lens structure remains unclear. Kormendy & Illingworth (1979) suggests that these lens features are a dissolving or partially dissolved bar. Like bars, some lenses may have their origin in/with rings. Each type of ring (see below) has a

[3] I would have preferred if we would have stuck with "ovals" since it describes what we see and when discussing "galaxy lenses" we typically now mean galaxy-size gravitational lenses.

lens analog (see Buta et al. 2015, and references therein). Or they could be something called a "barlens," vertically thick boxy/peanut section of the bar as viewed edge-on could look like a lens (Salo & Laurikainen 2017; Laurikainen & Salo 2017; Laurikainen et al. 2018).

4.6 Rings

Rings are prominent features in some galaxies. Often marked by recent star formation, rings may be completely closed or partial ("pseudo-rings"). Rings can be both narrow and sharp or broad and diffuse. Multiple rings in a galaxy disk are possible, ranging from nuclear rings to far into the disk. Rings are often linked to bars and can be explained as resonance features starting at the end of the bar or 2 × the bar radius. Once there is an asymmetry in the mass of a rotating disk, stars further out will get nudged into resonant orbits i.e., rings (Laurikainen et al. 2013; Comerón et al. 2014). Examples of resonance rings, thanks to a bar are nicely illustrated by NGC 3081 in Figure 4.14. However, the link between bars and rings is not always clear (Díaz-García et al. 2019).

Other origins for rings are accretion: new material has settled in a ring orbit around the center of a galaxy. This is thought to be the origin of polar rings, and collisional rings, the polar rings are perpendicular to the disk and collisional rings can have any angle to the disk.

Figure 4.14. NGC 3081, a nearly face-on spiral, barred galaxy. The galaxy's barred spiral center is surrounded by an inner resonance ring. This ring is full of bright clusters and bursts of new star formation, and surrounds the supermassive black hole. It is not unusual for rings like this to be seen in barred galaxies, as the bars are very effective at gathering gas into these resonance regions, causing pile-ups which leads to star formation. This can be seen here in both the inner ring as well the outer ring structures. Reprinted with permission from ESA/Hubble & NASA (R. Buta (University of Alabama)) from https://esahubble.org/images/potw1423a/.

There are multiple causes for rings (accretion, collision, resonances) and they result in different kinds of rings in the disk. There are nuclear rings, embedded in the central disk and bulge, rings halfway trough the disks and ones that seem to be tenuously linked to the central galaxy (Hoag's Object, Figure 4.15). Rings that appear perpendicular to the axes of the central galaxy (polar rings). Like many

Figure 4.15. Hoag's Object, a ring galaxy where the star-forming ring and central galaxy seem barely connected. Galaxies such as these occur after a bull's eye collision. Credit: NASA and the STScI.

morphological features, there are clues in the subtleties on their origin but there are radically different processes producing a similar looking phenomenon. As a rule of thumb, the rings that seem embedded in the disks of galaxies form mostly secularly, through processes like resonances of stellar orbits, and those at an angle to the plane of the galaxy come from external processes (collisions, tidal effects etc). Like all these morphological features traditionally classified by experts, these are now found by citizen scientist (cf Buta et al. 2017, and Chapter 5).

4.7 Lopsidedness

Galaxies are not completely axisymmetric, often we find the (outer) isophotes are elongated in one half of the galaxy (see for example M101 in Figure 7.19 or Figure 7.20).

Despite knowing about this for decades (Sandage 1961), astronomers tend to cheerfully ignore this fact and assume axisymmetric galaxies because it is much simpler to study the dynamics and photometry of axisymmetric disks.

Deviation from axi-symmetry is best observed where dust and other phenomena skew our view least; near-infrared and observations of the neutral hydrogen. The term "lopsided" was introduced (Baldwin et al. 1980) to highlight the non-axisymmetric nature of galaxy disks. This is slightly different from asymmetric or asymmetry discussed in Chapter 6.

A galaxy is characterized to be lopsided if it shows a global non-axisymmetric spatial distribution of type $m = 1$, where m is the azimuthal wavenumber, or a $\cos(\phi)$ distribution, where ϕ is the azimuthal angle in the plane of the disk.

Lopsidedness is one of those phenomena that stands out to the human eye but much less so in radial profiles (they smooth out lopsidedness by radially averaging) and really shows up in a Fourier decomposition of the image. While conceptually perhaps straightforward (take an image Fourier transform, mapping the dominant structural wavelengths to characterize typical scales on the image), the devil is, as often, in the details: poor foreground object subtraction results in too much weight in small scales, sky structure extra structure on large scales. Often is can be a good way to identify all the instrumental effects that happen on arcmin scales.

Figure 4.16 shows the 3.6 μm image from Spitzer and the different Fourier modes. $m = 2$ is where one should see the effect of lopsidedness (Rix & Zaritsky 1995) define A1, the fractional amplitude of the first azimuthal Fourier component ($m = 1$) of surface brightness, to be the quantitative measure of disk lopsidedness. A1 increases with radius. The average value measured between 1.5 and 2.5 disk scale lengths is large 0.1, and 30% of the galaxies studied show a higher lopsidedness (Rix & Zaritsky 1995). Figure 4.17 shows the radial dependence of the Fourier modes in the middle panel. The $m = 2$ mode shoots up mid-disk and takes over from $m = 1$, which is radial dependence. If you check out the radial profile, there are just some mild wobbles, easily explained away as "spiral structure or something." The "something" is lopsidedness.

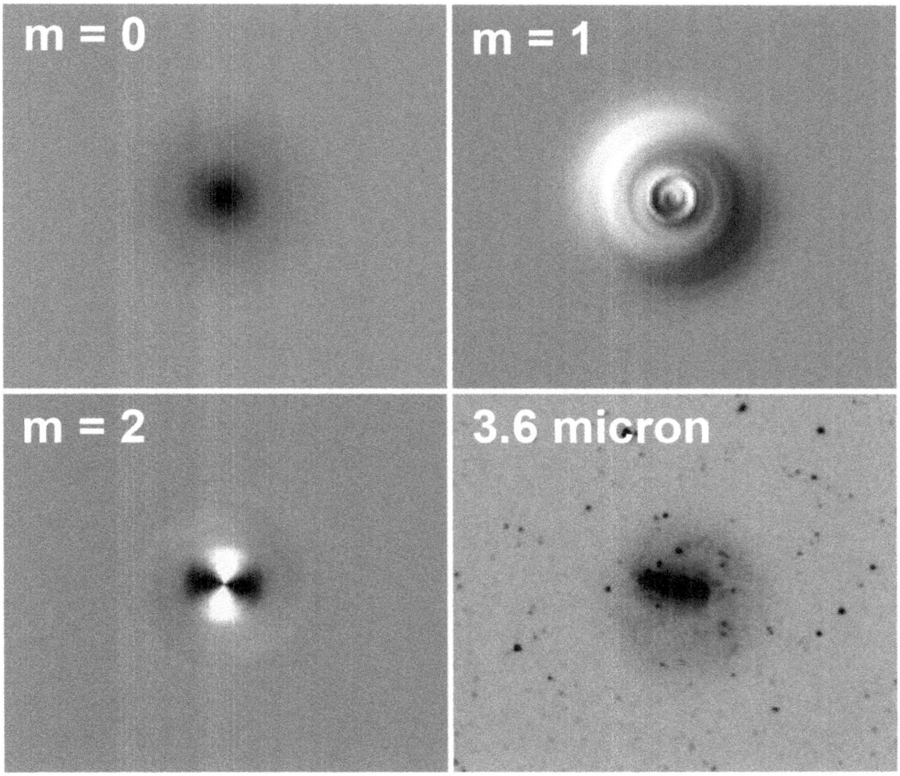

Figure 4.16. Decomposition of $m = 0$, 1, and 2 Fourier modes for NGC 3906 at arbitrary grayscales. The lower right panel shows the galaxy image at 3.6 μm. The image is 4′ on a side. $m = 2$ is where we expect lopsidedness. Image adapted from Zaritsky et al. (2013).

Lopsidedness is not only evident in the infrared imaging but also 21 cm atomic hydrogen line radio observations (HI) of face-on galaxies. The lopsided distribution of HI in M101 was noted in the early HI observations (Beale & Davies 1969; Baldwin et al. 1980; Noordermeer et al. 2001; Swaters et al. 2002; Noordermeer et al. 2005).

HI distributions were found to be quite commonly to be lopsided, with M101 still the textbook case (Figure 4.18). There was a lot of anecdotal evidence for spatial lopsidedness of HI disks of galaxies but now that uniform observations of larger samples are becoming available, these can be studied systematically (Swaters et al. 2002; Noordermeer et al. 2005).

One of the first systematic studies of morphological lopsidedness of HI disks (van Eymeren et al. 2011) found that lopsidedness is hard to link to other properties of the galaxies. Some HI disks show lopsidedness signal and some do not (Figure 4.19). Similarly, the lopsidedness is not exactly linked to galaxy morphometrics, even when measured on the same HI data (Holwerda et al. 2011; see Figure 4.20).

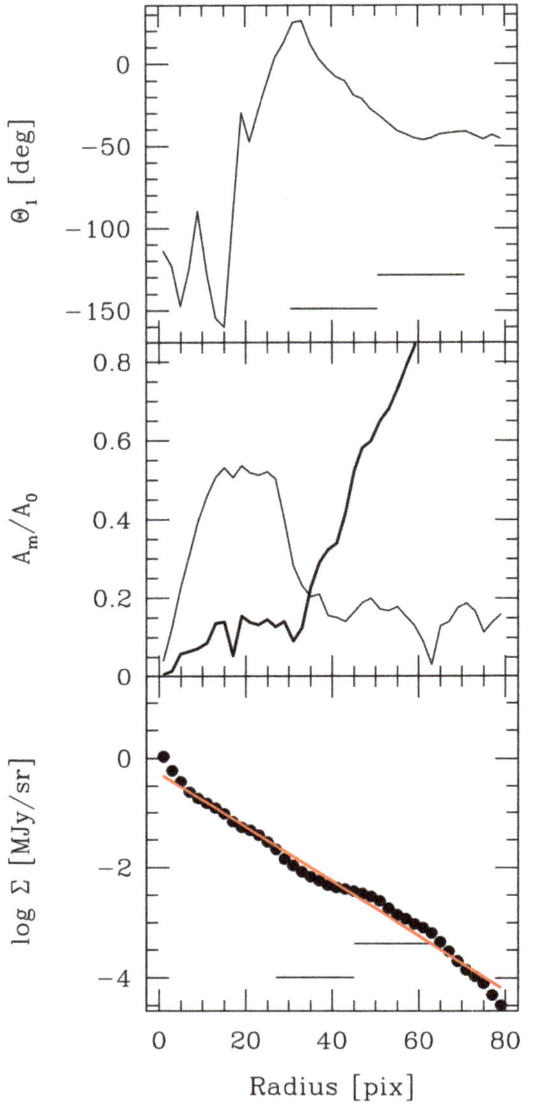

Figure 4.17. Fourier parameters for NGC 3906 (see Figure 4.16). Top panel: the $m = 1$ phase. Middle panel: the relative $m = 1$ amplitude (thick line) and $m = 2$ amplitude (thin line). Bottom panel: the surface brightness profile and fit. The $m = 1$ component is associated with lopsidedness in the outer radial interval. Image adapted from Zaritsky et al. (2013). © 2013. The American Astronomical Society. All rights reserved.

Lopsided morphology remains therefore difficult to link to other phenomena but in the case of HI is may point to gas accretion (Bournaud et al. 2005; Vulcani et al. 2018), or the effects of travel through the intergalactic medium; ram-pressure stripping.

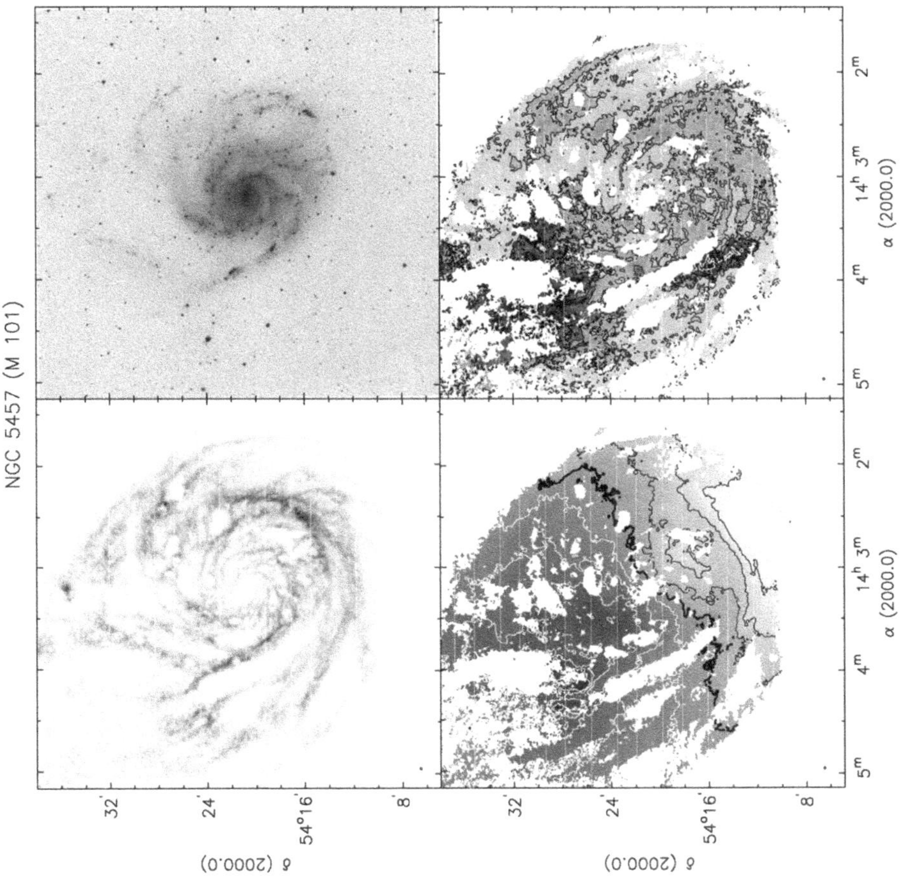

Figure 4.18. Top left: integrated HI map with grayscale ranging from 0 to 510 Jy km s $^{-1}$. Top right: optical image from the digitized sky survey (DSS). Bottom left: velocity field (moment 1). Black contours (lighter grayscale) indicate approaching emission, white contours (darker grayscale) receding emission. The thick black contour is the systemic velocity (v_{sys} = 226.4 km s $^{-1}$), the isovelocity contours are spaced bys Δv = 25 km s $^{-1}$. Bottom right: velocity dispersion map (moment 2). Contours are plotted at 5, 10, and 20 km s $^{-1}$ widths. The HI emission is clearly lopsided toward the North-West with kinematic lopsidedness evident as well. Figure reprinted from Walter et al. (2008).

Kinematic lopsidedness Lopsidedness is not just used for galaxy morphology but also its kinematics. Deviation from axi-symmetry in kinematics can be found in velocity maps and galaxy spectral profiles. The topic is treated in detail in Jog & Combes (2009) and van Eymeren et al. (2011) or a mix of morphometrics and lopsidedness in appearance and kinematics can be found in Reynolds et al. (2020).

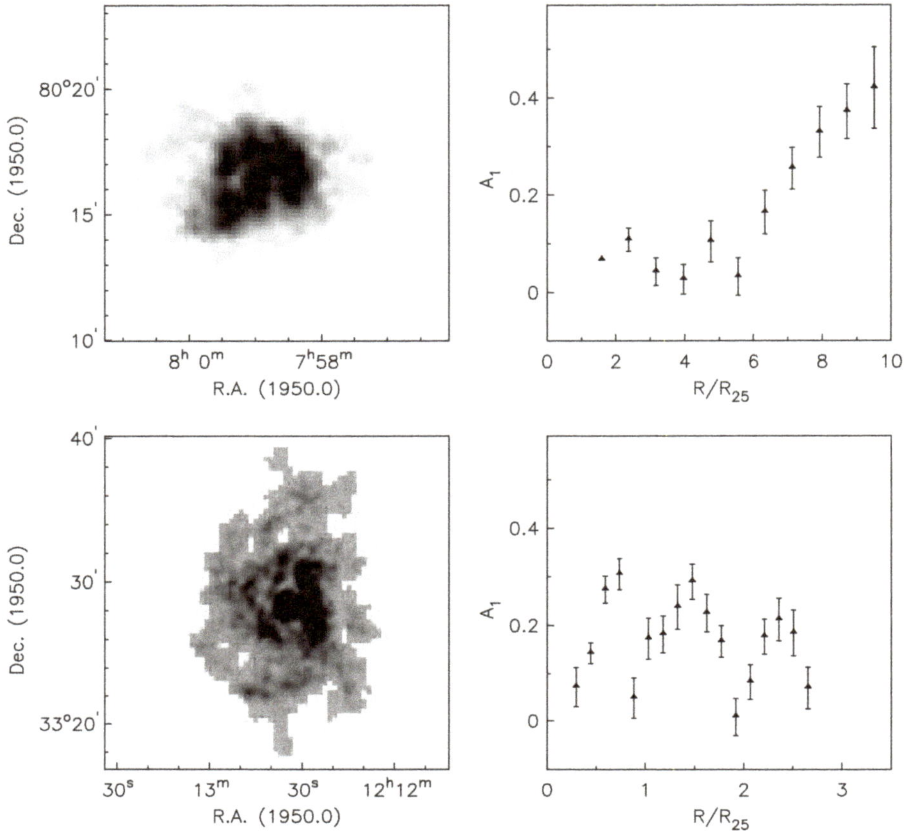

Figure 4.19. The HI intensity map and the distribution of A1 of two example galaxies, UGC4173 and UGC7256, one showing lopsidedness and an HI disk without. Figure adapted with permission from van Eymeren et al. (2011) © ESO.

4.8 Power Spectra

Instead of classifying all the various components in an image, one could take the power spectrum of the image to identify how much structure there is at a particular scale. The issue remains that one then has to have (a) removed foreground/background objects well, (b) sample on small enough scales to not be dominated by noise.

Figure 4.21 shows an example of an image power spectrum. Typical scales can be thought of as a wavelength (λ) so there are physical scales at parsec scales, tens of parsec scales (e.g., starclusters), kiloparsec scales (spiral arms, bars), and finally the galaxy as whole on tens of kiloparsec scales. The power spectrum is plotted in terms of wavenumber however $k = \frac{2\pi}{\lambda}$, and so large scales have a small wavenumber. Figure 4.21 shows the Herschel Space Telescope 100 micron image of M33. This wavelength is dominated not by star light but light from dust grains. So the structures we see are not stellar ones, like before but of the material in between the stars (interstellar matter or ISM). Some of these structures correspond to spiral arm

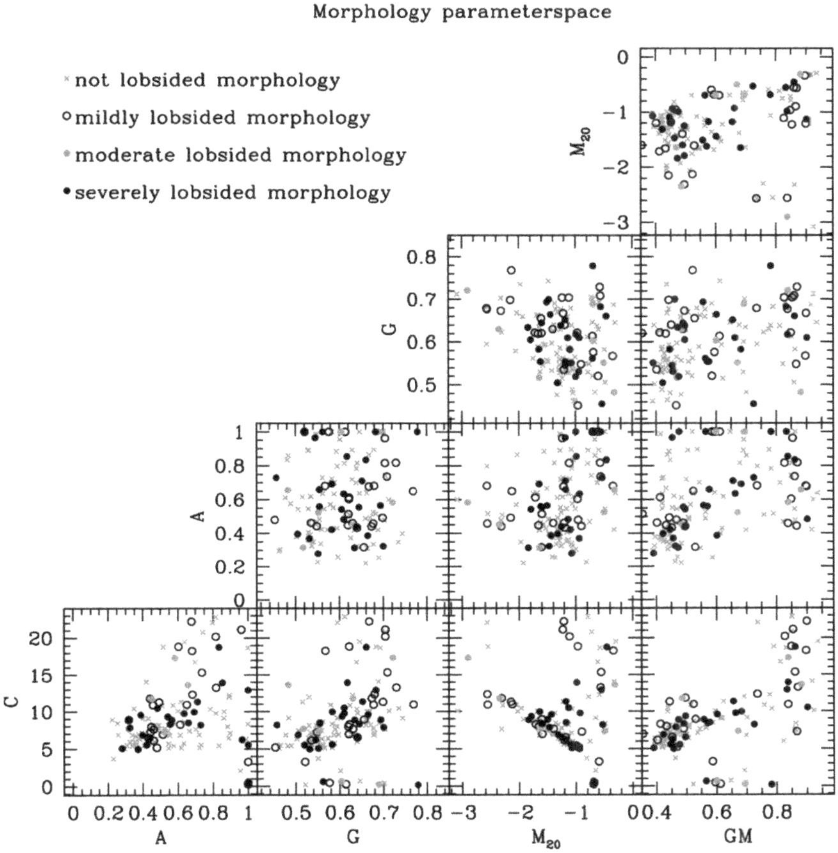

Figure 4.20. The distribution of morphometric parameters of the combined samples from Swaters et al. (2002) and Noordermeer et al. (2005) for the different morphologically lopsided classifications (none, mild, moderate, and severe). There is no clear preference of a lopsidedness classification for any part of the morphometric parameter space. Figure reproduced from Holwerda et al (2011). © 2011 The Authors Monthly Notices of the Royal Astronomical Society © 2011 RAS.

structure and some of the clouds correspond to HII regions or the birth site or starclusters. But the ISM is a gas mix and behaves as such with pressure and turbulence shaping it. Thus the large-scale structure corresponds well to what we see in stars (disk, arms) but the smaller scales are dominated by turbulence. The power spectrum's slope shows how strong the turbulence is.

4.9 Stellar Clusters

4.9.1 Globular Clusters

Globular clusters (GCs) are the spherical star groupings surrounding galaxies. There a number of them around our own Milky Way. Some of these originate from the earliest formation of the galaxy, and others are remains from other (smaller) galaxies accreted onto the main galaxy. The GC of the Milky Way offered a first clue that the

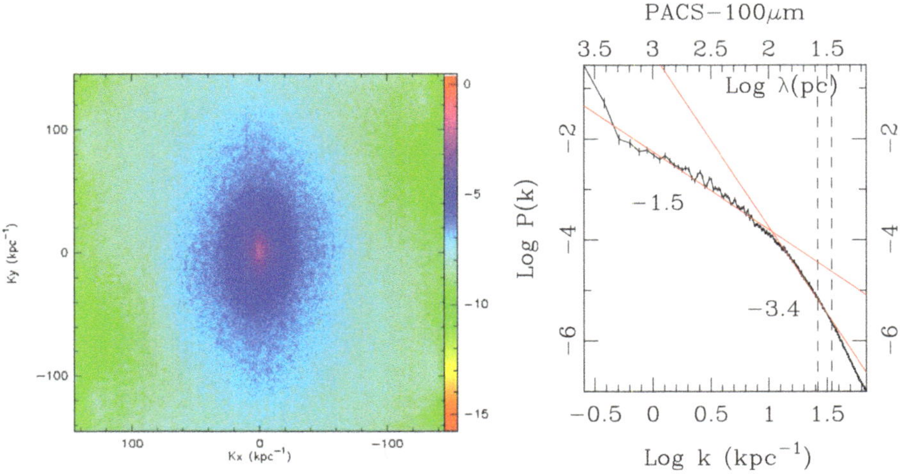

Figure 4.21. Left: the two-dimensional power spectrum of the Herschel (PACS camera) 100 micron image of M33, the Triangulum galaxy. Right: the power spectrum of this image. Low values of $\log k$ correspond to the larger scales in the image, the higher values to smaller scale. The wavelength (typical scale λ) is plot on the top axis. Two slopes can be fit to the two regimes in scales: the large-scale structure in the galaxy is described as $P(k) \propto -1.5 \times \log k$ and the small-scales as $P(k) \propto -3.4 \times \log k$. Especially this relation at the smaller scales hint at the prevalence of turbulence. Figure adapted with permission from Combes et al. (2012) © ESO.

Sun was not in the center of the Milky Way but well away from it: the GCs are radially distributed around the center, not the Sun.

The stars in a GC are thought to be all born at approximately the same and from the same material.[4] Because they are all made at the same time, the color–magnitude of GC reflect the state of of all stars at a particular age.

The population of GCs surrounding a disk galaxy is thought to originate (in part) from the same small galaxies that originally formed the stellar halo surrounding the massive central galaxy. The metallicities of their original formation cloud are reflected in the bimodal color distribution (Bailin & Harris 2009). Therefore, mapping the cluster population reveals the formation history of a galaxy, predominantly through the scaling relations for GC populations (Georgiev et al. 2010; Hudson et al. 2014).

4.9.2 Young Star Clusters and HII regions

Young star clusters are confined to the disks of galaxies and the site of the majority of their star formation. The bright clumps of young stars ionize the ISM around them resulting in HII[5] regions: clouds of glowing hydrogen, very visible in the emission lines of hydrogen and other atoms.

Once the output of ionizing ultraviolet starlight has abated (the most massive O-stars start to die off) and the ISM in the cluster is either used up or blown away by

[4] This is one of those 90% remarks. Yes broadly true but there is some evidence of two generations in GCs.
[5] HII is the description for Hα emission, the first in the Balmer series of hydrogen recombination lines.

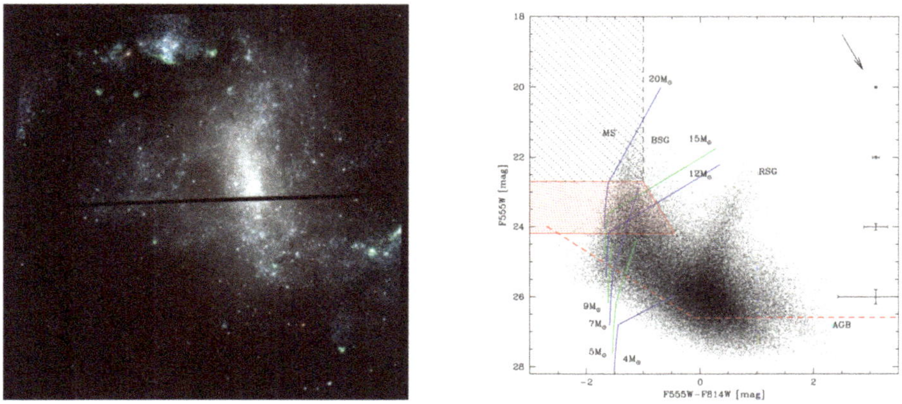

Figure 4.22. Left panel: the Hubble image of NGC 1313, a nearby dwarf spiral galaxy. Right panel: color–magnitude diagram of the resolved stars in NGC 1313 using the F555W (blue) and F814W (red) filters. Stellar evolutionary tracks for various masses shown in blue and green. The main sequence (MS), blue supergiant (BSG), red supergiant (RSG), and asymptotic giant (AGB) branches are marked. Figure reproduced from Pellerin et al. (2007). © 2007. The American Astronomical Society. All rights reserved.

stellar winds and supernovae, the remainder is a collection of stars in a tight grouping.

At that point, tidal effects on the cluster and star-star gravitational interactions work to dissolve the cluster into the disk at large. Figure 4.22 shows NGC 1313, a nearby, face-on dwarf disk galaxy. The youngest stars live on the main sequence with the most massive and brightest the shortest-living. All the brightest young stars are in clusters (Panel b in Figure 4.23). The next step in brightness main sequence stars tend to live longer but those are spread out more throughout the disk (panel c in Figure 4.23) and the dimmest main sequence stars live longer still and those are spread throughout the disk of NGC 1313.

What does that all mean? It means that as stars age, the diffuse throughout the disk: stellar clusters dissolve.[6]

The dissolution time of star clusters in the disk and how many stars are born in young (sometimes called "open") clusters remains a point of contention. Chandar et al. (2017) note that at every density of star formation, the fraction of stars in clusters holds steady at 25% no matter how intense the star formation is (Figure 4.24).

The Panchromatic Hubble Andromeda Treasury (PHAT; Dalcanton et al. 2012) survey observed a full third of the disk of the nearby Andromeda (M31) galaxy with the Hubble Space Telescope (Figure 4.25). This provides the best of both kinds of optical observations, a large overview of most of a galaxy, in this case Andromeda (M31), as well as resolved stars in the image.

[6] The nomenclature is "infant mortality" of star clusters, which is another example of astronomers naming things badly or gruesomely.

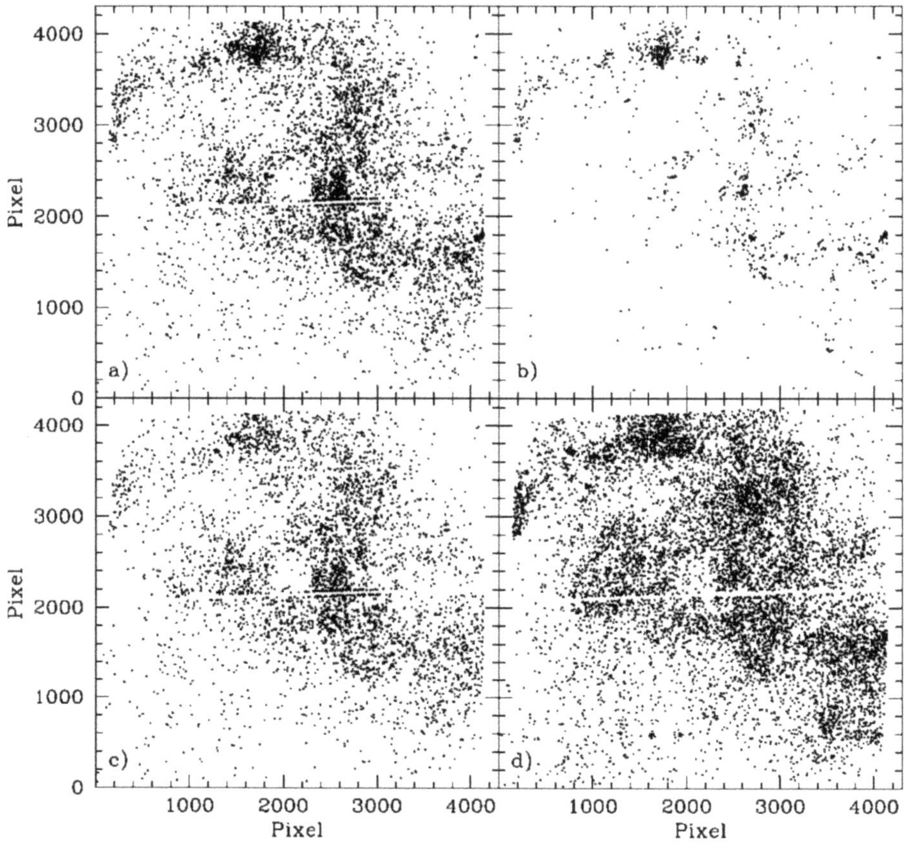

Figure 4.23. The stars in NGC 1313 divided by age: (a) all main sequence stars. (b) Youngest, most massive, and brightest stars in the upper region of the color–magnitude diagram (black shaded region in Figure 4.22). (c) Early B-type and late O-type stars within the lower region of the color–magnitude diagram (red shaded region in Figure 4.22). (d) Faintest stars located below the red shaded region of the color–magnitude diagram. Panel (b) are the shortest-lived stars, (c) are mid-lifetime O- and B-type stars and panel (d) are the faintest main sequence stars (F-type stars) expected to live much longer. The longer stars live, the further away they can be found from stellar clusters: many star clusters must therefore dissolve into the disk. Figure reproduced from Pellerin et al. (2007). © 2007. The American Astronomical Society. All rights reserved.

This means each individual cluster of blue, young stars can be identified (for example by citizen scientists, more on that in Chapter 5) and studied both as a population of objects (i.e., cluster) and as collections of individual stars. Figure 4.26 shows the detail revealed by HST resolution on stellar clusters: a sea of stars in the disk with dense blue clumps of young stars, just emerged from the birth cloud, see Figure 4.27.

4.9.3 Nuclear Star Clusters

Nuclear star cluster are bright clumps of stars in the very center of galaxies (hence the nuclear) and occur in both spiral and elliptical galaxies. They stand out especially in bulge-less or low prominent-bulge disk galaxies. These central objects

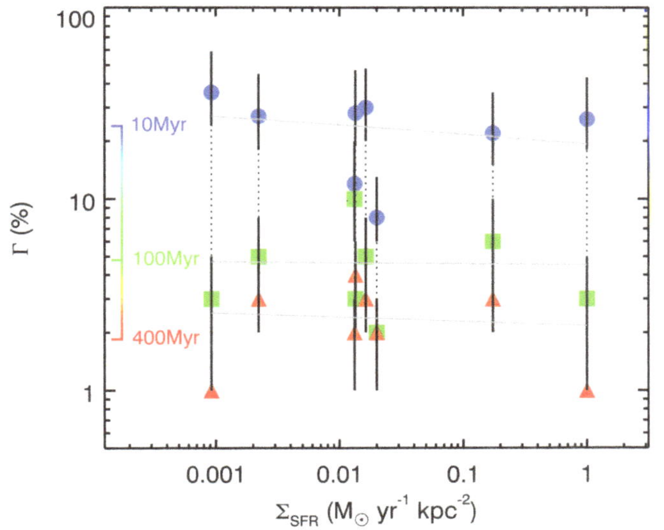

Figure 4.24. The fraction of stars in clusters at formation (10 Myr, blue), after 100 Myr (green) and 400 Myr. 1 Myr is a million years. Figure reproduced from Chandar et al. (2017). © 2017. The American Astronomical Society. All rights reserved.

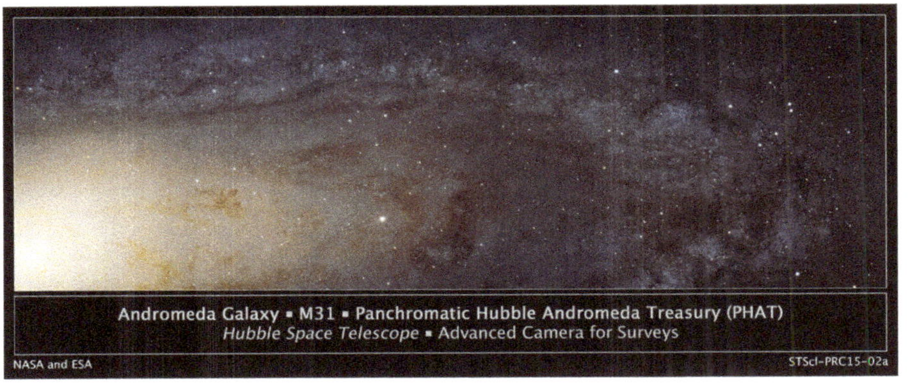

Figure 4.25. The largest mosaic ever made with the Hubble Space Telescope: one third of the disk of Andromeda for the Panchromatic Hubble Andromeda Treasury (PHAT) project. Both cameras onboard Hubble took hundreds of images to be stitched together in the multi-color mosaic. A 120 million individual stars can be found in this image. The majority is in the disk of Andromeda but many star clusters can be found (Figure 4.26). Credit: NASA, ESA, J. Dalcanton, B.F. Williams, and L.C. Johnson (University of Washington, the PHAT team, and R. Gendler, from https://hubblesite.org/images/gallery/73-phat.

are inevitably linked to the supermassive black hole supposedly at the center of most galaxies. Do these cluster evolve with them or are they a product of their presence?

The previous two types of clusters are—by and large—the product of local conditions but nuclear star cluster properties seem to be tied somewhat to the properties of the host galaxy, very much like the properties of host galaxies and supermassive black holes. Figure 4.28 shows how more massive nuclear clusters are

Figure 4.26. Left: a detail of the large Hubble mosaic in Figure 4.25. Several star clusters are highlighted and zoom-ins are shown in the right panels. Each cluster is made up of young, blue stars clustered together. With Hubble one can resolve these in Andromeda and figure out their ages. Credit: NASA, ESA, J. Dalcanton, B.F. Williams, and L.C. Johnson (University of Washington, the PHAT team, and R. Gendler, from https://hubblesite.org/contents/media/images/2015/18/3552-Image.html?Topic=105-galaxies&keyword=M31.

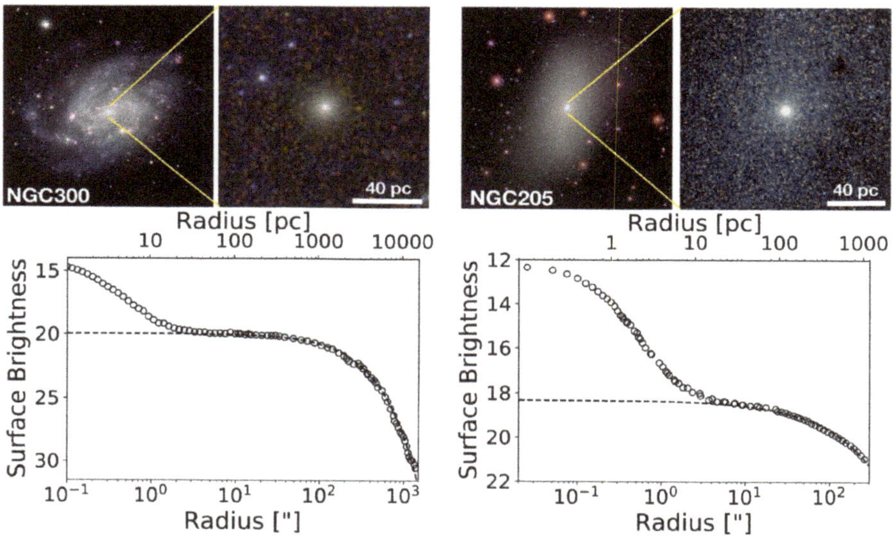

Figure 4.27. The Nuclear Star Cluster in the late-type spiral NGC300 (left) and early-type galaxy NGC205 (M110, right). The top panels show galaxy-wide images with zoom-ins into the central regions of each galaxy. The bottom panels show the inner surface brightness profiles of the two galaxies. In both cases indicate the presence of a nuclear star cluster as a rise above the light profile of the host galaxy body (points are data, dashed line fits; Böker et al. 2002; Kim et al. 2004; Bland-Hawthorn et al. 2005; Valluri et al. 2005). The large-scale image of NGC300 is from https://www.eso.org/public/images/eso1037a/, while the inset HST image is from Carson et al. (2015). © 2015. The American Astronomical Society. All rights reserved. The large-scale image of NGC205 is from Sloan imaging (retrieved from https://cseligman.com/text/atlas/ngc2.htm, image credit: the Sloan Digital Sky Survey), while the inset HST image is from https://www.nasa.gov/feature/goddard/2018/messier-110. Composite figure reprinted with permission from Springer Nature: from Neumayer et al. (2020), Copyright © 2020, The Author(s)

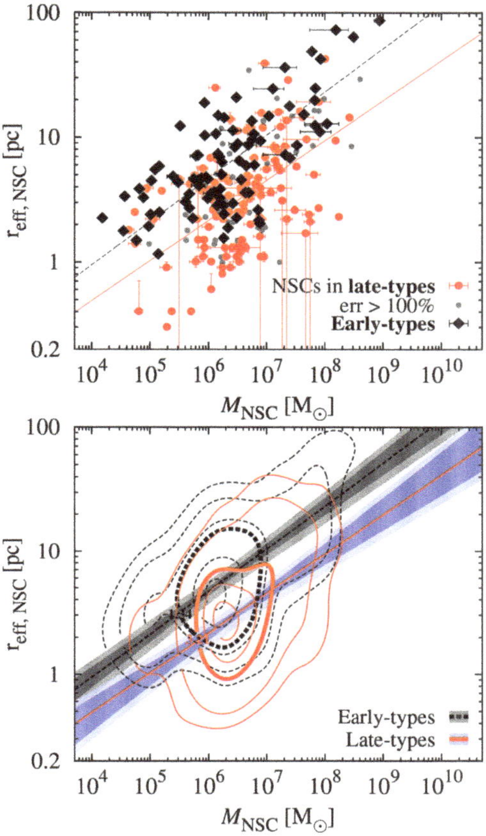

Figure 4.28. The size–mass relation of nuclear star clusters: the effective radius as a function of estimated stellar mass. Noteworthy is how the relation is different for spirals (red) as opposed to ellipticals (black points). Nuclear star cluster properties do depend on the type of galaxy they reside in. Figure reproduced from Georgiev et al. (2016). © 2016 The Authors, Published by Oxford University Press on behalf of the Royal Astronomical Society.

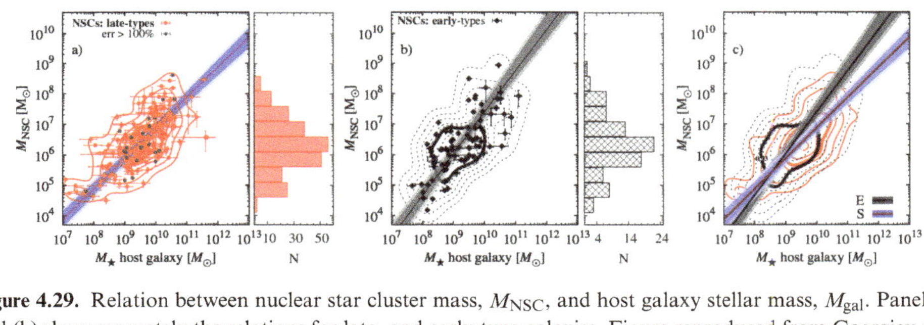

Figure 4.29. Relation between nuclear star cluster mass, M_{NSC}, and host galaxy stellar mass, M_{gal}. Panels (a) and (b) show separately the relations for late- and early-type galaxies. Figure reproduced from Georgiev et al. (2016). © 2016 The Authors, Published by Oxford University Press on behalf of the Royal Astronomical Society.

also bigger (not so much a surprise) but also that how mass and size are related depends on what kind of galaxy these nuclear star clusters are in.

Their mass depends on the host galaxy (again, bigger things are part of bigger things) but that relationship also depends on the type of galaxy! (see Figure 4.29). Correlations between the central star cluster and the properties of the galaxy at large make these interesting clusters to study.

4.10 Dust Lanes

Dust lanes are particular morphological features as (in the optical) they are in essence absences of light. In color images, they end up looking like brown/red streaks across the disk of stars. Because of their nature and the fractal appearance (they look like rivers of murk self similar down to the resolution of your images), they are often not sub-classified and their presence noted. In the edge-on perspective of massive disk galaxies, we do expect the cumulative effect of dust clouds along our line-of-sight to result in a dark band mid-plane. For this reason, dust lanes are a good way to identify perfectly edge-on galaxies (e.g., NGC 891) and a few degrees off (e.g., NGC 5907).

4.11 Assignments

There are two jupyter notebooks that can go with this chapter: Bulge Types (Jupyter notebook 4) and the spiral arm assignment (Jupyter notebook 5).

4.12 Further Reading

Galaxy morphological components are described in great detail in Buta with detailed explanations (Buta 2011), application to Spitzer data (Buta et al. 2015) and a comparison with Galaxy Zoo (Buta et al. 2017). Ron Buta also has an excellent book on galaxy Morphology.

Bulges: Wyse et al. (1997) and Renzini (1999) for reviews on the classical bulges. For an in-depth review of pseudo-bulges (Kormendy & Kennicutt 2004) is an excellent review. For the details on how to model the shapes of bulges inside disks, see Mendez-Abreu (2016).

Lopsidedness: the topic of lopsidedness is treated in exhaustive detail in Jog & Combes' review article (Jog & Combes 2009) or a recent update on the state of things in Zaritsky et al. (2013).

Nuclear Star Cluster: the overview by Neumayer et al. (2020) comes highly recommended.

References

Abraham, R. G., Merrifield, M. R., Ellis, R. S., Tanvir, N. R., & Brinchmann, J. 1999, MNRAS, 308, 569

Andredakis, Y. C., Peletier, R. F., & Balcells, M. 1995, MNRAS, 275, 874

Andredakis, Y. C., & Sanders, R. H. 1994, MNRAS, 267, 283

Athanassoula, L. 2003, in Lecture Notes in Physics, Vol. 626, Galaxies and Chaos, ed. G. Contopoulos, N. Voglis (Berlin: Springer), 313

Bailin, J., & Harris, W. E. 2009, ApJ, 695, 1082

Baldwin, J. E., Lynden-Bell, D., & Sancisi, R. 1980, MNRAS, 193, 313
Beale, J. S., & Davies, R. D. 1969, Natur, 221, 531
Binney, J., & Tremaine, S. 1987, Galactic dynamics (Princeton, NJ: Princeton Univ. Press)
Bland-Hawthorn, J., Vlajic, M., Freeman, K. C., & Draine, B. T. 2005, ApJ, 629, 239
Böker, T., Laine, S., van der Marel, R. P., et al. 2002, AJ, 123, 1389
Bournaud, F., & Combes, F. 2002, A&A, 392, 83
Bournaud, F., Combes, F., Jog, C. J., & Puerari, I. 2005, A&A, 438, 507
Buta, R., Sheth, K., Athanassoula, E., et al. 2011, ApJS, 217, 32
Buta, R. J. 2011, arXiv:1102.0550
Buta, R., et al. 2017, MNRAS, 471, 4027
Buta, R. J., Knapen, J. H., Elmegreen, B. G., et al. 2009, AJ, 137, 4487
Carson, D. J., Barth, A. J., Seth, A. C., et al. 2015, AJ, 149, 170
Chandar, R., Fall, S. M., Whitmore, B. C., & Mulia, A. J. 2017, ApJ, 849, 128
Churchwell, E., Babler, B. L., Meade, M. R., et al. 2009, PASP, 121, 213
Collier, A., Shloshman, I., & Heller, C. 2018, MNRAS, 476, 1331
Comerón, S., Elmegreen, B. G., Salo, H., et al. 2014, A&A, 571, A58
Combes, F., Boquien, M., Kramer, C., et al. 2012, A&A, 539, A67
Dalcanton, J. J., Williams, B. F., Lang, D., et al. 2012, ApJS, 200, 18
Díaz-García, S., Diaz-Suarez, S., Knapen, J. H., & Salo, H. 2019, A&A, 625, A146
Díaz-García, S., Salo, H., & Laurikainen, E. 2016, A&A, 596, A84
Du, M., Ho, L. C., Debattista, V. P., et al. 2020, ApJ, 895, 139
Elmegreen, B. G., Elmegreen, D. M., & Hirst, A. C. 2004, ApJ, 612, 191
Elmegreen, D. M., & Elmegreen, B. G. 1987, ApJ, 314, 3
Elmegreen, D. M., Elmegreen, B. G., Yau, A., et al. 2011, ApJ, 737, 32
Erwin, P., Saglia, R. P., Fabricius, M., et al. 2015, MNRAS, 446, 4039
Fisher, D. B. 2006, in ASP Conf. Ser. 352, New Horizons in Astronomy: Frank N. Bash Symposium, ed. S. J. Kannappan, S. Redfield, J. E. Kessler-Silacci, M. Landriau, & N. Drory (San Francisco, CA: ASP), 237
Fisher, D. B., & Drory, N. 2008, AJ, 136, 773
Gadotti, D. A. 2009, MNRAS, 393, 1531
Gao, H., Ho, L. C., Barth, A. J., & Li, Z.-Y. 2020, ApJS, 247, 20
Georgiev, I. Y., Böker, T., Leigh, N., Lützgendorf, N., & Neumayer, N. 2016, MNRAS, 457, 2122
Georgiev, I. Y., Puzia, T. H., Goudfrooij, P., & Hilker, M. 2010, MNRAS, 406, 1967
Hart, R. E., Bamford, S. P., Hayes, W. B., et al. 2017, MNRAS, 472, 2263
Holwerda, B. W., Pirzkal, N., de Blok, W. J. G., et al. 2011, MNRAS, 416, 2415
Hudson, M. J., Harris, G. L., & Harris, W. E. 2014, ApJL, 787, L5
Jog, C. J., & Combes, F. 2009, PhR, 471, 75
Jogee, S., Barazza, F. D., Rix, H.-W., et al. 2004, ApJL, 615, L105
Kennicutt, R. J., & Hodge, P. 1982, ApJ, 253, 101
Kennicutt, R. C. J. 1981, AJ, 86, 1847
Kim, S. C., Sung, H., Park, H. S., & Sung, E.-C. 2004, ChJAA, 4, 299
Knapen, J. H. 2010, in Galaxies and their Masks, 201
Kormendy, J. 1993, in IAU Symp. 153, Galactic Bulges (Kluwer: Academic Publishers), 209
Kormendy, J., & Illingworth, G. 1979, in Photometry, Kinematics and Dynamics of Galaxies (Austin, TX: Univ. of Texas at Austin), 195
Kormendy, J., & Kennicutt, R. C. Jr 2004, ARA&A, 42, 603

Krajnović, D., Karick, A. M., Davies, R. L., et al. 2013, MNRAS, 433, 2812
Laurikainen, E., & Salo, H. 2017, A&A, 598, A10
Laurikainen, E., Salo, H., Athanassoula, E., et al. 2013, MNRAS, 430, 3489
Laurikainen, E., Salo, H., Laine, J., & Janz, J. 2018, A&A, 618, A34
Lin, C. C., & Shu, F. H. 1964, ApJ, 140, 646
Masters, K. L., Lintott, C. J., Hart, R. E., et al. 2019, MNRAS, 487, 1808
Mendez-Abreu, J. 2016, in Astrophysics and Space Science Library, Vol. 418, Galactic Bulges (Cham: Springer), 15
Méndez-Abreu, J., Simonneau, E., Aguerri, J. A. L., & Corsini, E. M. 2010, A&A, 521, A71
Neumayer, N., Seth, A., & Boker, T. 2020, A&ARv, 28, 4
Noordermeer, E., Sparke, L. S., & Levine, S. E. 2001, MNRAS, 328, 1064
Noordermeer, E., van der Hulst, J. M., Sancisi, R., Swaters, R. A., & van Albada, T. S. 2005, A&A, 442, 137
Pellerin, A., Meyer, M., Harris, J., & Calzetti, D. 2007, ApJL, 658, L87
Reid, M. J., Menten, K. M., Brunthaler, A., et al. 2019, ApJ, 885, 131
Reid, M. J., Menten, K. M., Zheng, X. W., et al. 2009, ApJ, 700, 137
Reissl, S., Stil, J. M., Chen, E., et al. 2020, A&A, 642, A201
Renzini, A. 1999, Ap&SS, 267, 357
Reynolds, T. N., Westmeier, T., Staveley-Smith, L., Chauhan, G., & Lagos, C. D. P. 2020, MNRAS, 493, 5089
Ringermacher, H. I., & Mead, L. R. 2009, MNRAS, 397, 164
Rix, H-W, & Zaritsky, D. 1995, ApJ, 447, 82
Salo, H., & Laurikainen, E. 2017, AAS/DAM 48, 303.02
Salo, H., Laurikainen, E., Buta, R., & Knapen, J. H. 2010, ApJL, 715, L56
Sandage, A. 1961, The Hubble Atlas of Galaxies
Schulze, F., Remus, R.-S., Dolag, K., et al. 2018, MNRAS, 480, 4636
Seigar, M. S., Bullock, J. S., Barth, A. J., & Ho, L. C. 2006, ApJ, 645, 1012
Seigar, M. S., Davis, B. L., Berrier, J., & Kennefick, D. 2014, ApJ, 795, 90
Seigar, M. S., & James, P. A. 1998a, MNRAS, 299, 672
Seigar, M. S., & James, P. A. 1998b, MNRAS, 299, 685
Sellwood, J. A., & Wilkinson, A. 1993, RPPh, 56, 173
Sheth, K., Elmegreen, D. M., Elmegreen, B. G., et al. 2008, ApJ, 675, 1141
Simard, L., Mendel, J. T., Patton, D. R., Ellison, S. L., & McConnachie, A. W. 2011, ApJS, 196, 11
Swaters, R. A., van Albada, T. S., van der Hulst, J. M., & Sancisi, R. 2002, A&A, 390, 829
Valluri, M., Farrarese, L., Merritt, D., & Joseph, C. L. 2005, ApJ, 628, 137
van Eymeren, J., Jutte, E., Jog, C. J., Stein, Y., & Dettmar, R. J. 2011, A&A, 530, A29
Vulcani, B., Poggianti, B. M., Moretti, A., et al. 2018, ApJ, 852, 94
Walmsley, M., Lintott, C., Geron, T., et al. 2022, MNRAS, 509, 3966
Walter, F., Brinks, E., de Blok, W. J. G., et al. 2008, AJ, 136, 2563
Wyse, R. F. G., Gilmore, G., & Franx, M. 1997, ARA&A, 35, 637
Xu, H., Zhu, L., Grand, R., et al. 2019, MNRAS, 489, 842
Xu, Y., Bian, S. B., Reid, M. J., et al. 2018, A&A, 616, L15
Zaritsky, D., Salo, H., Laurikainen, E., et al. 2013, ApJ, 772, 135
Zhang, X., & Buta, R. J. 2007, AJ, 133, 2584
Zhu, L., van den Bosch, R., van de Ven, G., et al. 2018, MNRAS, 473, 3000

Chapter 5

Visual Classifications

Visual classification is the oldest form of morphology quantization. In Chapter 1, we talked about visual classifications and there is an excellent overview by Buta (2011), so why another chapter?

Because we are no longer leaving it to the experts.

The issue with visual classification (apart from the fact that it is laborious and a tad repetitive) is that it is *subjective* and that is a problem. Even when classified by a team of people who trust each other's eye, there is typically a subclass of spread in these classifications: sometimes they all agree, sometimes they do not. These classifications needed an error i.e., a confidence range.

Enter citizen science, specifically the Galaxy Zoo project (Lintott et al. 2008, 2011). In this case the classification is broken up in a series of unambiguous questions (Figure 5.1) that can be asked from anyone on the internet, and then ask them to anyone who is interested in helping science. Now for a given question like "is this galaxy smooth?" there are statistics on what classifiers voted for e.g., 80% of people seem to think so.

Note that even though we now have taken the individual classifier out of the equation (with his or hers unambiguous weight; you either trust their judgment or you do not) we now have a *crowd*. And as any sociologist can tell you, you do not necessarily make the problem simpler by adding more people. But this is data, and data can be calibrated. Against an expert, by weighting votes of "experienced" Galaxy Zoo users higher. And it reveals where the questions, so carefully worded, can be in fact be open to interpretation once shown to many people. A good example in Figure 5.1 is the "Is there anything odd?" That is a value judgment. Like the "Irr" category of the Hubble Tuning Fork, it is a catch-all. A necessity but it does require everyone to think the same things as "odd"[1] when classifying galaxies. This was later rectified in further Galaxy Zoo classifications.

[1] This galaxy isn't odd, you're odd.

doi:10.1088/2514-3433/ac2c7dch5

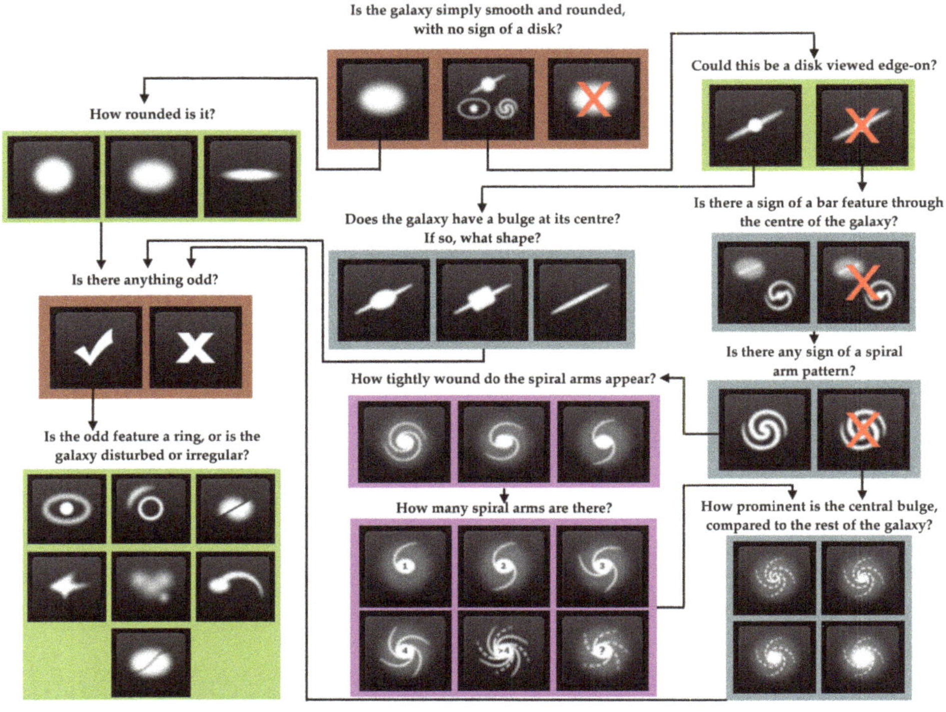

Figure 5.1. The decision flowchart of the classification tasks for GalaxyZoo.v2 (GZ2), beginning at the top center. Tasks are color-coded by their relative depths in the decision tree. Tasks outlined in brown are asked of every galaxy. Tasks outlined in green, blue, and purple are (respectively) one, two, or three steps below branching points in the decision tree. Figure reproduced from Willett et al. (2013). © 2013 The Authors, Published by Oxford University Press on behalf of the Royal Astronomical Society.

And showing this to a crowd has unexpected benefits. Without some of the prejudices that come from an astronomical educations, the volunteers were free to ask what something they saw was, see Figures 5.2 and 5.3.

Hanny's "Voorwerp"[2]

In 2007 while classifying galaxies on the Galaxy Zoo website, citizen scientist Hanny van Arkel came across an object that she was unsure how to classify. It appeared as blue squiggle (for want of a better word) near the spiral galaxy IC 2497. She flagged it, and further investigation revealed that it was at the same distance as IC 2497 and comparable in size to the Milky Way. It was a brand new type of astronomical object previously unknown to scientists, and discovered by a citizen!

How does a Dutch school teacher become interested in classifying galaxies on Galaxy Zoo? Because she likes the band Queen. And Queen's guitar player Brian May is also an astrophysicist, promoting this project on social media.

[2] "Voorwerp" is Dutch for "object." So when astronomers now talk about the "voorwerp"' class of objects, they are talking about "object objects," like "Sahara desert" or "Lake Tahoe" etc.

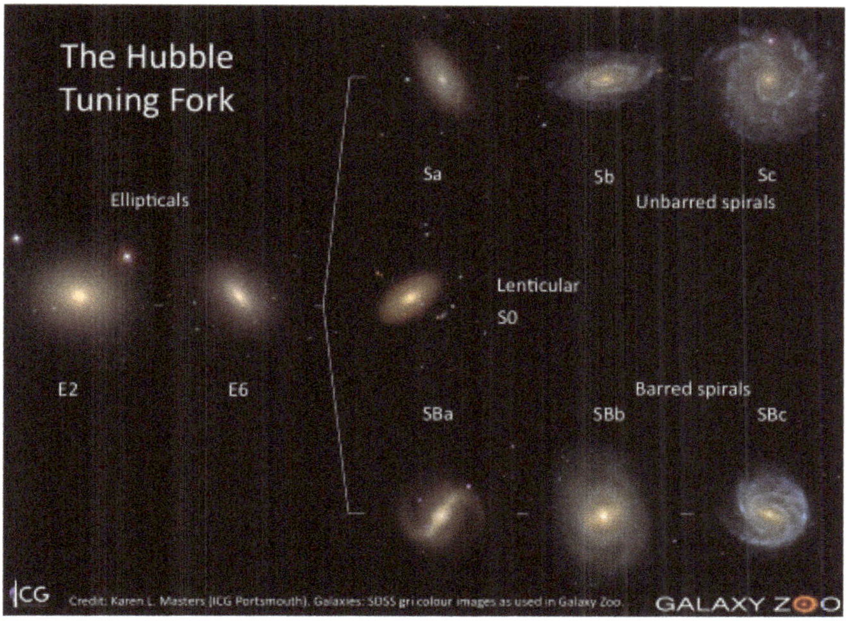

Figure 5.2. A new version of the Hubble tuning fork with Galaxy Zoo examples. Figure originally by Karen Masters. By combining the right questions in Figure 5.1, one can assign a Hubble type to each galaxy. Credit: the Galaxy Zoo Team (https://www.zooniverse.org/projects/zookeeper/galaxy-zoo)

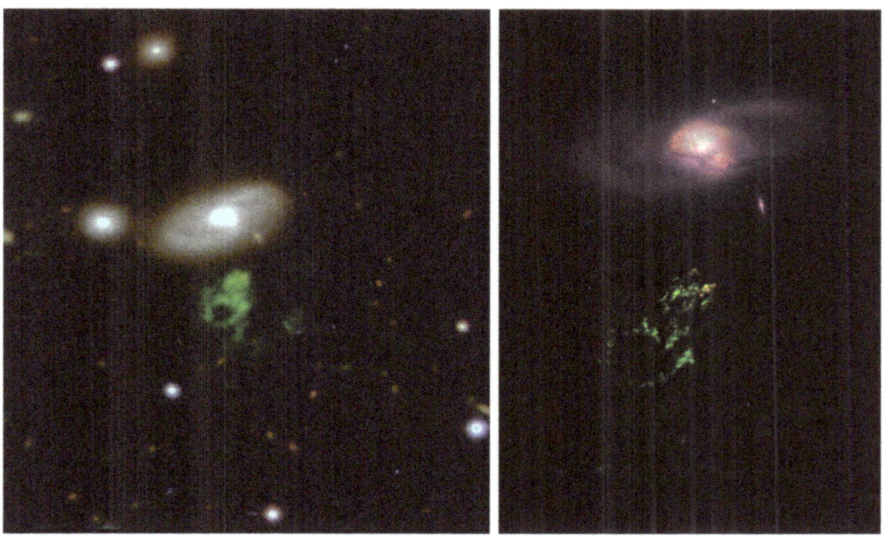

Figure 5.3. Two views of Hanny's Voorwerp, the mysterious cloud of glowing oxygen near IC 2497 (the galaxy above it). A ground-based image using the William Herschel Telescope (left) and a Hubble Space Telescope image (right). The enormous green cloud is suspected to be part of a tidal tail of material illuminated by the quasar (supermassive black hole shining beams of high energy light and particles from its poles) inhabiting the center of IC 2497. The quasar suddenly turned off, leaving only galaxy and the glowing "voorwerp" visible in telescopes at optical wavelengths. Figures from NASA's Astronomy Picture of the day. Credit: (left) the Galaxy Zoo Team (https://www.zooniverse.org/projects/zookeeper/galaxy-zoo), (right) NASA, ESA, W. Keel (University of Alabama), and the Galaxy Zoo Team.

5.1 Personal Example

Another example of crowd sourcing science is the Andromeda Project. This was a visual inspection of every part of the Panchromatic Andromeda Treasury Program, a Hubble Space Telescope project to image 1/3 of the nearby Andromeda galaxy (Figure 4.25). This is a huge Hubble color mosaic with millions of stars. The goal was to identify where those stars lived close together in star clusters and how many of those clusters there were (Figure 4.26). Computers are notoriously bad at this. A major source of contamination is background galaxies seen through the disk of Andromeda. For my PhD, my project had been to count the background galaxies in Hubble images to determine how transparent the foreground galaxies actually were. I counted a lot of them, some 2500. It took me a year (I had to calibrate, make sure to minimize biases etc). The Andromeda project also counted background galaxies (in the process of counting clusters). They found 2700+. It took them three weeks.

The lesson for graduate students and other professional astronomers: at some point someone will effectively redo your thesis. It will cost an almost comically little amount of effort. This is progress. It's ok.

Second lesson: the really quick classification was only possible by a massive upfront investment by the Andromeda Project team to set up and test the citizen science machinery. With citizen science, you can get this right only once. If the project is buggy, people will leave and never come back.

When to do a citizen science project?

The zooniverse infrastructure allows any scientist to set up and start promoting a citizen science project. But when is it good to do that? The Galaxy Zoo people have a rule of thumb:

Hundreds of objects: do it yourself. Time to suck it up Princess Buttercup.[3] Thousands of objects: make it a zooniverse project but just do the classifications with you and your immediate collaborators. 10,000 or more classifications needed: citizen science is definitely faster.

5.2 Special Features

Once a questionnaire like the one in Figure 5.1 has been decided on, the answers from Galaxy Zoo (or a group of astronomers) can be compared to traditional measures of galaxy properties, e.g., color or luminosity or Sérsic index.

One can divide galaxies in blue (star-forming), green (intermediate or quenching), and red (quiescent) and then evaluate how often certain features are identified (Kelvin et al. 2018). Figure 5.4 shows examples of the same features (bar, ring, lens, tidal tails, shells, and interaction) in red, green, and blue galaxies.

[3] Look I know none of you have watched "The Princess Bride" alright? Do yourself a favor and watch it. It holds up.

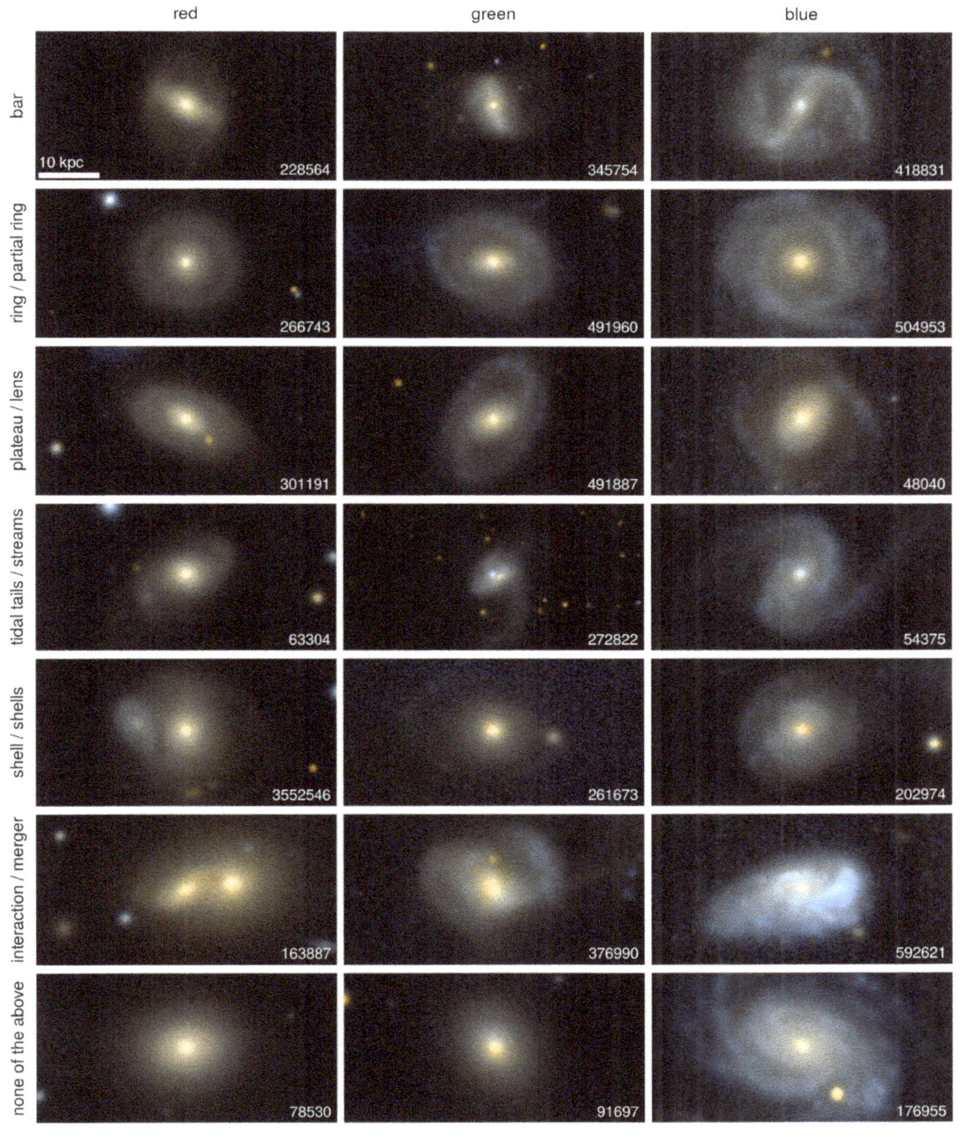

Figure 5.4. Examples for each of the six structural components classified (top six rows), separated into red, green, and blue, as indicated in each column. The bottom row shows example galaxies in which none of the six structural markers were identified. Postage stamps are constructed using KiDS g and r band imaging, arcsinh scaled. Each galaxy is shown at the same physical scale as represented by the scale bar in the top left panel. Figure reproduced from Kelvin et al. (2018). © 2018 The Author(s), Published by Oxford University Press on behalf of the Royal Astronomical Society.

With voting tallies, one can infer *uncertainties* in a morphological classification. Figure 5.5 shows what a group classification can show: the green galaxies stand out in their number of lenses and rings. All the other features are consistent with gradients or the same fraction across galaxy populations. Does the same mechanism

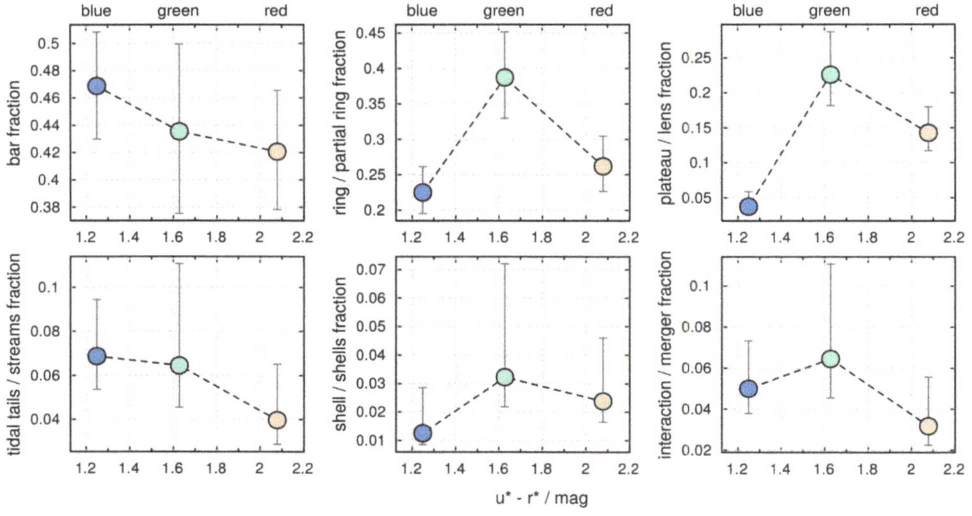

Figure 5.5. The voting fractions for various features: bars, rings, lens, tidal streams, shells, and merger remnant as a function of $u - r$ color (blue–green–red) for disk-type galaxies. Note that rings and lenses appear significantly more often in green galaxies. Does the same mechanism that makes a ring or a lens, suppress the star formation? Or does suppressed star formation induce these features? Figure reproduced from Kelvin et al. (2018). © 2018 The Author(s), Published by Oxford University Press on behalf of the Royal Astronomical Society.

that causes a ring or lens morphology in a disk galaxy also quench its star formation, moving it into the green galaxy category? Or does reduced star formation open the possibility for ring and lens structures to form?

5.3 Radio Zoo

A more recent development has been an increase in images in radio wavelengths. Radio images often have suffered from both noise and the fact that they lack colors. Radio images often lack the splendor of optical astronomy, despite the fact they contain so much new information. But that has not stopped the Radio Zoo team and volunteers. Identifying radio galaxy morphologies and how the angle of the jets are related to each other is something a dedicated group of people can do.

Radio Galaxy Zoo focused on a couple topics like every iteration of the Galaxy Zoo has: identifying which galaxy the radio continuum emission belonged to (Figure 5.6), and then characterizing the radio emission by, for example, measuring the angles between the radio jets and the galaxy orientation (Figure 5.7).

5.4 Clump Scout

Galaxy Zoo has been steadily increasing the range of data classified and expanding what is asked of citizen scientists. A more recent addition is Galaxy Zoo: Clump Scout. In this project, the fine scale structure, the star clusters where new stars are formed are the focus. Figure 5.8 illustrates the specific kinds of galaxies this project is looking for. The classification problem remains of comparable size to Galaxy Zoo

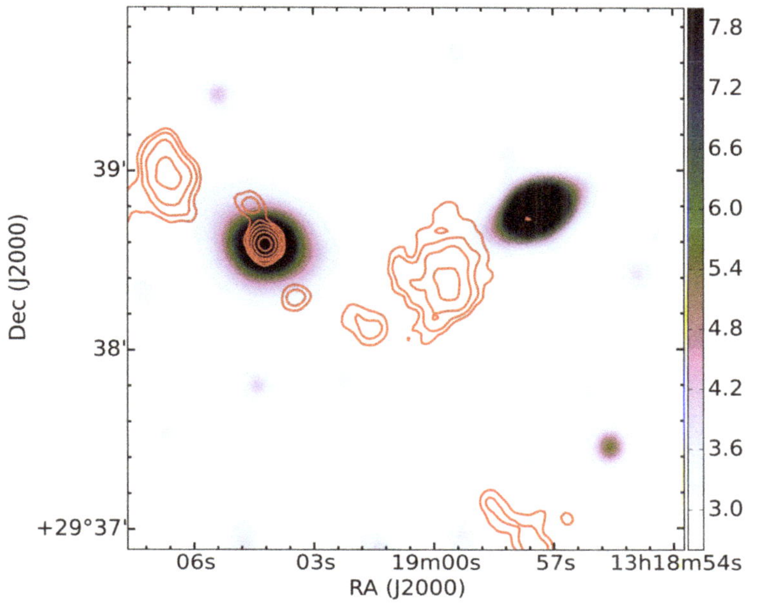

Figure 5.6. An example of a galaxy where visual identification of the radio components is critical. Automated algorithms would have classified the non-core emission as independent sources, whereas Radio Galaxy Zoo volunteers (in agreement with the science team) find all five radio emission components in the upper half of the image to be related to the same source. Figure reproduced from Banfield et al. (2015). © 2015 The Authors, Published by Oxford University Press on behalf of the Royal Astronomical Society.

but the focus has specified into smallest scale structure. Other projects have gone to identify spiral arm structure or take a closer look of galaxies in the green valley or explore XUV disks. The Zoo is far from done doing science.

5.5 Machine-aided Crowds

In the near future (see Chapter 11), new observatories will take images of millions of galaxies. This may prove too ambitious for citizen science projects. Keeping in mind that every image ideally is classified 100 or so times, one would need a sizeable fraction of the online world population to participate.

Enter machine-assisted citizen science. Like various photo apps, the machine-assisted citizen science project finds objects very similar to the one that is presented to the citizen scientists for inspection, once there is quorum on this object (90% think this is an elliptical), the machine assistance attaches that classification to the similar objects as well. Figure 5.9 shows how much faster such a machine assist speeds up classification: the dashed line is how fast object retires objects to be classified. Machine learning quickly learns similar objects, groups them with the same classification and retires them. The comparative Galaxy Zoo 2 project is shown as well. Speeding classifications up (and keeping it less monotonous) shows a viable way forward for citizen science.

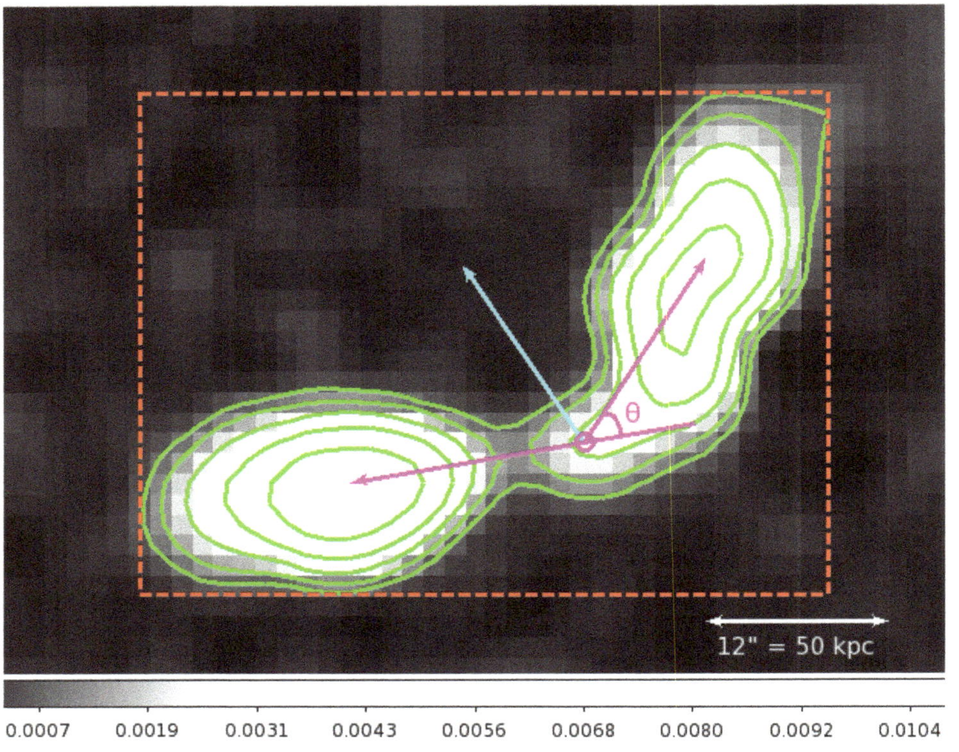

Figure 5.7. Example source RGZ J080 641.4+494629. The magenta arrows extend from the host galaxy identified by RGZ users and terminate at the peaks of the radio emission, defining the bending angle θ. The cyan arrow is used to define the orientation for the galaxy (position angle) with respect to the cluster. Figure reproduced from Garon et al. (2019).

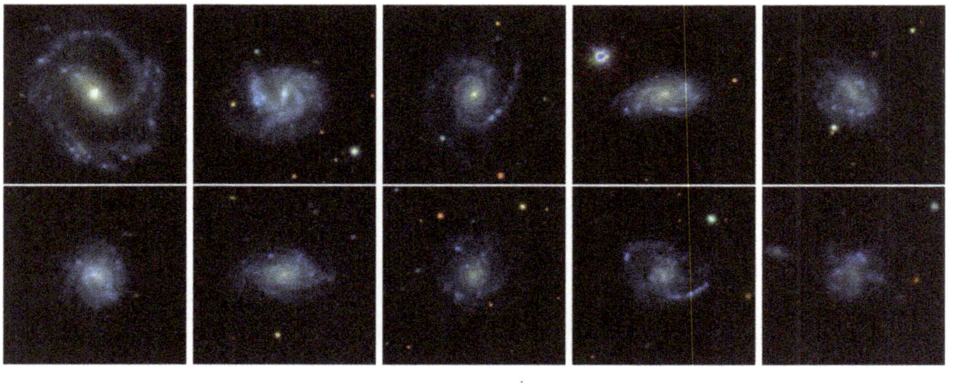

Figure 5.8. A small sample of clumpy galaxies found by the Galaxy Zoo: Clump Scout project. These are some of the galaxies identified to have clumps of star formation throughout their disk. Figure Credit: the Galaxy Zoo Team (https://www.zooniverse.org/projects/zookeeper/galaxy-zoo).

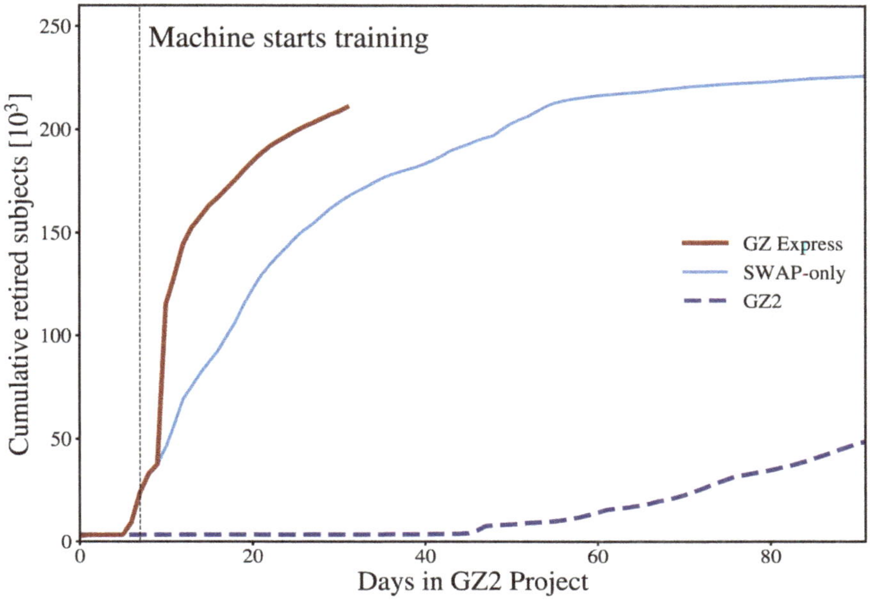

Figure 5.9. By incorporating a machine classifier, GZX (red) increases the classification rate by an order of magnitude compared to GZ2 (dashed dark blue) and out-performs the SWAP-only run (light blue) retiring more than 200K subjects in just 27 days of GZ2 project time. The dashed black line marks the first night the machine trains. After several additional nights of training, it is deemed optimized and allowed to retire subjects. Both humans and machine then contribute to retirement. The simulation after 32 days having retired over 210K galaxies (the size of GZ2). Figure reproduced from Beck et al. (2018). © 2018 The Author(s), Published by Oxford University Press on behalf of the Royal Astronomical Society.

Mind you, machines help you find more of the same things. For the truly unique, a human remains competitive (see the inset about Hanny' voorwerp).

5.6 Assignments

The data from Galaxy Zoo is available at https://data.galaxyzoo.org. Each science paper from the first few Galaxy Zoo projects are there in machine readable format.

There are two Galaxy Zoo jupyter notebooks to go with this chapter: GalaxyZoo-1 (Jupyter notebook 8) and GalaxyZoo-2 (Jupyter notebook 9).

5.7 Further Reading

GalaxyZoo-1 (Lintott et al. 2008, 2011) or GalaxyZoo-2 (Willett et al. 2013; Hart et al. 2016) describe the process of Galaxy Zoo in great detail. The Galaxy Zoo papers by Willett et al. (2011, 2013, 2017) are comprehensive in detail about the Galaxy Zoo data and results.

References

Banfield, J. K., Wong, O. I., Willett, K. W., et al. 2005, MNRAS, 453, 2326
Beck, M. R., Scarlata, C., Fortson, L. F., et al. 2018, MNRAS, 476, 5516

Buta, R. J. 2011, arXiv:1102.0550
Garon, A. F., Rudnick, L., Wong, O. I., et al. 2019, AJ, 157, 126
Hart, R. E., Bamford, S. P., Willett, K. W., et al. 2016, MNRAS, 461, 3663
Kelvin, L. S., Bremer, M. N., James, P. A., et al. 2018, MNRAS, 477, 4116
Lintott, C., Schawinski, K., Bamford, S., et al. 2011, MNRAS, 410, 166
Lintott, C. J., Schawinski, K., Sloasr, A., et al. 2008, MNRAS, 389, 1179
Willett, K. W., Darling, J., Spoon, H. W. W., et al. 2011, ApJ, 730, 56
Willett, K. W., Galloway, M. A., Bamford, S. P., et al. 2017, MNRAS, 464, 4176
Willett, K. W., Lintott, C. J., Bamford, S. P., et al. 2013, MNRAS, 435, 2835

Chapter 6

Galaxy Morphometrics

A completely different observational approach to characterizing galaxy appearances is "non-parametric" parameters. In essence, several authors and groups have tried to come up with parameters that can be applied to images that do not depend too much on resolution or a preconceived idea about the shape of the profile.[1] The idea is that you can easily compute these on catalogs of thousand to millions of objects without having to specify fit ranges or other manual input. These non-parametric or "morphometric" parameters[2] can then be used as quantities to classify galaxies along the Hubble tuning fork or to identify mergers in a population of galaxies.

The morphometric parameters discussed are Concentration, Asymmetry and Smoothness from Conselice (2003), M_{20} and Gini from Lotz et al. (2004), and the MID parameters from Peth et al. (2016) and Rodriguez-Gomez et al. (2019).

In many cases, these morphometrics still require input other than the image. The relevant input parameters are the central position of the galaxy (x_c, y_c), and a definition of the area over which these parameters are computed. The Gini parameter only requires the definition of the area and not the central position, making it less sensitive to input error.

6.1 CAS

CAS refers to the now commonly used Concentration–Asymmetry–Smoothness space (Conselice 2003) for morphological analysis of distant galaxies. Concentration of the light, symmetry around the center, and smoothness as an indication of substructure (see Figure 6.1).

[1] The words "too much" are doing an awful lot of work here.

[2] A term first coined here (Davenport 2015). The term "morphometrics" originates from biology to describe animal appearance in qualitative form. Think "length of fins" or how far eyes are set apart. Or the point system for deer.

doi:10.1088/2514-3433/ac2c7dch6

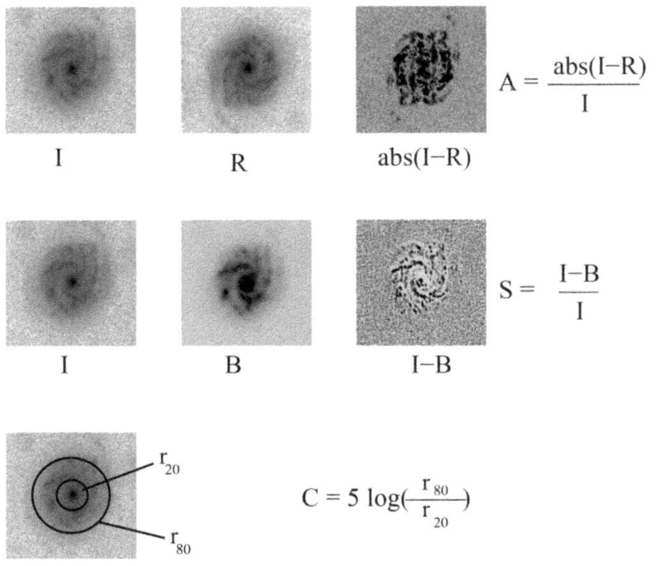

Figure 6.1. Graphical representation of how the CAS parameters are defined: asymmetry (A), clumpiness (S), and concentration (C). For the measurements of A and S, I is the original galaxy image, R is this image rotated by 180°, while B is the image after it has been smoothed by the factor $0.3 \times r(\eta = 0.2)$. See also Conselice et al. (2000) and Bershady et al. (2000). Figure reproduced from Conselice (2003).

Concentration is defined by Bershady et al. (2000) as:

$$C = 5 \log(r_{80}/r_{20}) \tag{6.1}$$

with r_{f} as the radius containing percentage f of the light of the galaxy (see definitions of r_f in Bertin & Arnouts 1996, Holwerda 2005).

The asymmetry is defined as the level of *point-*, (or rotational-) symmetry around the center of the galaxy (Abraham et al. 1994; Conselice 2003):

$$A = \frac{\Sigma_{i,j}|I(i, j) - I_{180}(i, j)|}{\Sigma_{i,j}|I(i, j)|}, \tag{6.2}$$

where $I(i, j)$ is the value of the pixel at the position i, j in the image, and $I_{180}(i, j)$ is the pixel at position $[i, j]$ in the galaxy's image, after it was rotated 180° around the center of the galaxy (Figure 6.1).

Inspired by the "unsharp masking" technique (Malin 1978), Smoothness is defined by Takamiya (1999) and Conselice (2003) as:

$$S = \frac{\Sigma_{i,j}|I(i, j) - I_S(i, j)|}{\Sigma_{i,j}|I(i, j)|} \tag{6.3}$$

where $I_S(i, j)$ is the same pixel in a smoothed image. What type of smoothing is used has changed over the years. Often a fixed Gaussian smoothing kernel is chosen for simplicity (Figure 6.1). The fact that this parameter has another input parameter in the form of the size of the smoothing kernel, makes it highly "tunable," meaning you get out exactly what you optimized for but it is very difficult to reliably compare between catalogs or even implementations.

6.2 Gini and M_{20}

Abraham et al. (2003) and Lotz et al. (2004) introduce the Gini parameter to quantify the distribution of flux over the pixels in an image. They use the following definition:

$$G = \frac{1}{\bar{I}n(n-1)}\Sigma_i(2i - n - 1)I_i, \tag{6.4}$$

I_i is the value of pixel i in an ordered list of the pixels, n is the number of pixels in the image, and $\bar{I}$ is the mean pixel value in the image (Figure 6.2). This is the computationally least expensive implementation. The Gini parameter is an indication of equality in a distribution (initially used as an economic indicator), still in use with the United Nations (Gini 1912; Yitzhaki 1991),[3] with G=0 the perfect equality

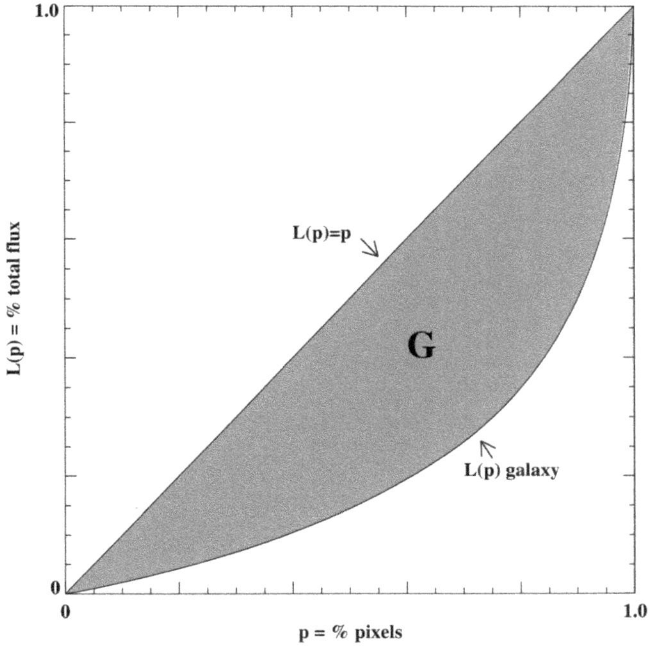

Figure 6.2. Lorenz curve: the Gini coefficient is the area between the Lorenz curve of the galaxy's pixels and that of equitable distribution (shaded region). The given curve is for S0 NGC 4526, $G = 0.59$. Equitable distribution of flux (or wealth) is when every pixel has the same fraction (straight line). Gini describes the deviation from equatable. Figure reproduced from Lotz et al. (2004). © 2004. The American Astronomical Society. All rights reserved.

[3] Gini had fascist sympathies but this is a good mathematical model to boil income inequality down to a single value.

(all pixels have the same fraction of the flux),[4] and $G = 1$ perfect inequality (all the intensity is in a single pixel).[5] Its behavior is therefore in between that of a structural measure and concentration. Gini appears quite sturdy as it does not require the center of the object to be computed. It remains relatively unchanged, even when the object is lensed (Florian et al. 2016). It remains popular for this reason. However, it depends strongly on the image's signal-to-noise (Lisker 2008). In essence, noise can add pixels with no fraction of the flux in them, artificially increasing the Gini value (Figure 6.2).

Lotz et al. (2004) also introduced a new way to parameterize the extent of the light in a galaxy image. They define the spatial second order moment as the product of the intensity with the square of the projected distance to the center of the galaxy. This gives more weight to emission further out in the disk. It is sensitive to substructures such as spiral arms and star-forming regions but insensitive if these are distributed symmetrically or not.

The second order moment of a pixel i is defined as:

$$M_i = I_i \times [(x - x_c)^2 + (y - y_c)^2], \tag{6.5}$$

where $[x, y]$ is the position of a pixel with intensity value I_i in the image and $[x_c, y_c]$ is the central pixel position of the galaxy in the H I surface density map.

The total second order moment of the image is given by:

$$M_{tot} = \Sigma_i M_i = \Sigma I_i[(x_i - x_c)^2 + (y_i - y_c)^2]. \tag{6.6}$$

Lotz et al. (2004) use the relative contribution of the brightest 20% of the pixels to the second order moment as a measure of disturbance of a galaxy:

$$M_{20} = \log\left(\frac{\Sigma_i M_i}{M_{tot}}\right), \quad \text{for } \Sigma_i I_i < 0.2 I_{tot}. \tag{6.7}$$

The M_{20} parameter is sensitive to bright regions in the outskirts of disks and thus higher values can be expected in galaxy images (in the optical and UV) with star-forming outer regions as well as those images of strongly interacting disks.

6.3 Gini Second Order Moment

Instead of using the intensity of pixels, we can define a Gini parameter for the second order moment of each pixel by substituting M_i (Equation (6.5)) for I_i in Equation (6.4):

$$G_M = \frac{1}{\bar{M} n(n-1)} \Sigma_i (2i - n - 1) M_i. \tag{6.8}$$

This is the contribution to the galaxy morphometric parameter space to provide an additional handle to characterize lopsidedness and interaction level (Holwerda

[4] Think the Smurfs in terms of economic equity.

[5] Or in economics, your garden variety dictatorship.

et al. 2011a, 2011b, 2011c, 2011d, 2011e). The reasoning was that the Gini parameter has the added benefit of using the combined shape of the flux distribution curve (all the information in the image), rather than just a fraction. M_{20} may be not sensitive enough to interaction signature while Asymmetry is sensitive to other effects as well (Holwerda et al. 2011b). A similar conclusion was reached by Lotz et al. (2008), hence $G_{\rm M}$ is an attempt to define a single parameter to detect interaction using all the information on the second order moment.

6.4 Patchiness

A recent addition to the morphometric parameter space is a "patchiness" parameter defines as:

$$P = -\ln\left\{\Pi_i^N \frac{1}{\sqrt{2\pi\sigma_i}} \exp\left[-\frac{(X_i - \bar{X}_w)^2}{2\sigma_i^2}\right]\right\} \tag{6.9}$$

where N_i is the number of pixels, X_i is the value of pixel i. The Gaussian probability that all the pixels equal the weighted average. This is lower when "patchier." The touted benefits are that this measure is sensitive to deviations above and below the average. Benefits, like the Gini parameter are that this parameter does not depend on the central position, unlike M_{20}, which relies on a bright subset of pixels and the object's central position (Fetherolf et al. 2020a, 2020b). At present it is not implemented in STATMORPH but that is something that is relatively easily remedied.[6]

6.5 MID Statistics

The MID statistics (Freeman et al. 2013; Peth et al. 2016) were introduced as an alternative to the Gini–M_{20} and CAS statistics to be more sensitive to recent mergers. However, these new statistics have not been tested as extensively as the Gini–M_{20} and CAS statistics, especially using hydrodynamic simulations (Lotz et al. 2008, 2010a, 2011; Bignone et al. 2017), see also the discussion in the implementation in STATMORPH (Rodriguez-Gomez et al. 2019).

6.5.1 Multimode (M)

The multimode statistic (M) measures the ratio between the areas of the two most "prominent" clumps within a galaxy (Figure 6.3). The implicit assumption is that this galaxy is well-resolved enough to even *have* two clumps. That is true for nearby galaxies from the ground and most galaxies with Hubble, James Webb, or Euclid/Roman. Its calculation mostly consists in finding such substructures. First, all pixels within the MID segmentation map are sorted by brightness. Then, for a given quintile q (between 0 and 1), the set of all pixels with flux values above the qth quintile will generally consist of n groups of contiguous pixels, which are sorted by area (largest first): $A_{q,1}$, $A_{q,2}$, ⋯, $A_{q,n}$. Finally, M is defined as the quintile q that maximizes the area ratio between the two largest groups (Peth et al. 2016):

[6] It is left as an exercise to the reader. I do have a snippet in my STATMORPH.

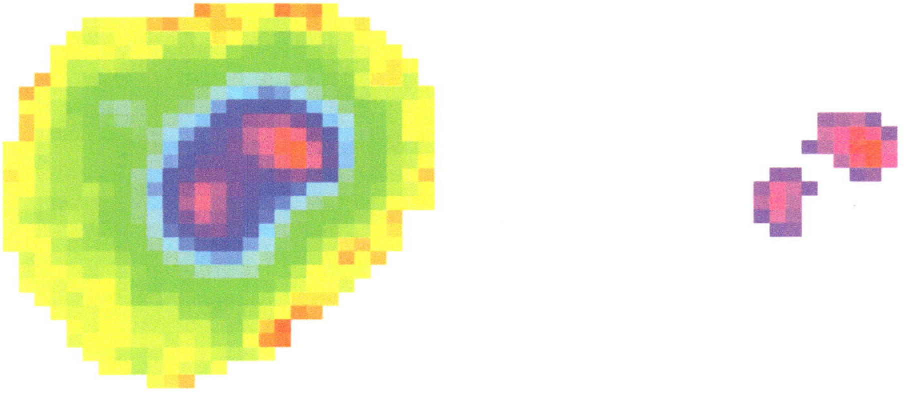

Figure 6.3. Example of pixel grouping for computing the multimode (M) statistic. To the left the H-band pixel intensities for an example of a merger galaxy, while to the right only those pixels associated with the largest 5.8 per cent of the sorted intensity values. The two pixel regions have areas $A(1) = 21$ and $A(2) = 14$ pixels, so that $R_{0.942} = (14/21) \times 14 = 9.33$. The M statistic is the largest of the R values computed for a sufficiently large number of threshold percentiles. Figure reproduced from Freeman et al. (2013). © 2013 The Authors, Published by Oxford University Press on behalf of the Royal Astronomical Society.

$$M = \max\left(\frac{A_{q,2}}{A_{q,1}}\right) \tag{6.10}$$

where $A_{q,1}$ is the largest quintile and $A_{q,2}$ is the second-to-largest quintile area. The issue remains that the segmentation image here had to come from *somewhere* and the decisions which peaks are separate objects and which belong to an objects have effectively been made there, in the construction of this map.

6.5.2 Intensity (I)

The intensity statistic (I) measures the ratio between the two brightest subregions of a galaxy (Figure 6.4). To calculate it, the galaxy image is first slightly smoothed using a Gaussian kernel with $\sigma = 1$ pixel. Then, the image is partitioned into pixel groups according to the watershed algorithm: each distinct subregion consists of all the pixels such that their maximum gradient paths lead to the same local maximum. Once the pixel groups are defined, their summed intensities are sorted into descending order: $I1$, $I2$, etc. The intensity statistic is then defined as (Freeman et al. 2013):

$$I = \frac{I_2}{I_1}. \tag{6.11}$$

The same issue that can be raised for M can be raised here. There is a built-in assumption of resolved structure and that this structure has not fractured the segmentation map into separate catalog entries.

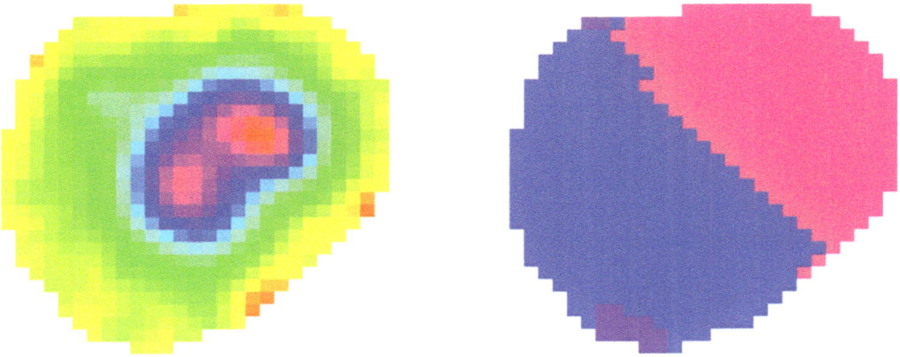

Figure 6.4. Example of pixel grouping for computing the intensity (I) statistic. To the left the pixel intensities for an example of a merger galaxy (see Figure 6.3). These data are smoothed using a symmetric Gaussian kernel with a width of $\sigma = 1$ pixel, a sufficiently small scale to remove local intensity maxima caused by noise without removing local maxima intrinsic to the galaxy itself. To the right pixel regions associated with each local intensity maxima which remain after smoothing. Pixel intensities are summed within each region, with the intensity statistic then being the ratio of the second-largest to largest sum. In this example, the statistic is $I = 0.935$. Figure reproduced from Freeman et al. (2013). © 2013 The Authors, Published by Oxford University Press on behalf of the Royal Astronomical Society.

6.5.3 Deviation (D)

The deviation statistic (D) measures the distance between the image centroid, (x_c, y_c), calculated for the pixels identified by the MID segmentation map, and the brightest peak found during the computation of the I statistic, (x_I, y_I). This distance is normalized by $\sqrt{n_{seg}/\pi}$, where n_{seg} is the number of pixels in the segmentation map, which represents an approximate galaxy "radius" (Freeman et al. 2013):

$$D = \sqrt{\frac{\pi}{n_{seg}}}\sqrt{(x_c - x_I)^2 + (y_c - y_I)^2}\,. \tag{6.12}$$

This is a metric that can be calculated from SE output by using the image centroid and the peak location, albeit again the this makes assumptions on the number of substructure in the image.

6.6 Implementation

The easiest application of these morphometrics is the STATMORPH code by Rodriguez-Gomez et al. (2019). This builds on segmentation maps (maps identifying which pixel belongs to which object in the catalog) made by source extractor (Holwerda 2005; Bertin & Arnouts 1996). If you want a reasonably quick morphometrics catalog, this is a good place to start.

6.7 Uncertainties

Most of the earliest implementations of these parameters came with no error estimate at all. In most cases because it was unclear how to compute it or it was a moot point because it would be relatively large.

Uncertainty estimates can be obtained from a bootstrapping run, varying the central position of each galaxy or randomly subsampling the pixel values in the galaxy-area (jacknifing). Key here is to use the same information (the object's pixels), randomize an input parameter (central position or smoothing kernel) or drop a subsample just to make sure your morphometric measurement is not dominated by a single hot pixel. The standard deviation of a dozen or so iteration of the morphometrics is a reasonable estimate of the error on each morphometric parameter (cf Holwerda et al. 2014).

6.8 Caveats

Caveat emptor and these morphometrics are not magical. For one thing, their values depend on the signal-to-noise (s/n) levels of the image itself. For example the Gini value changes with s/n of the image (Lisker 2008). This can be understood intuitively. If the sky level is noisier (lower s/n), more noisy pixels or sky pixels will be added to the object. These do not contribute any signal but artificially change the Gini values. In the case of HI morphometrics, Giese et al. (2016) note that a high signal-to-noise is critical for use of Asymmetry etc. Similarly Lotz et al. (2006) caution against trusting any morphometric derived for an object with a s/n < 2.5. As you may have noted in my description of the MID morphometrics, a lot depend on the segmentation map; which pixels are included in each object. This depends not only on the s/n but also how aggressive the object segmentation (subdivision into objects) is by the catalog software. This is always a compromise between removing foreground and unassociated objects and shattering a galaxy into its smaller substructures.

The most important caveat is that morphometrics can only be compared if they are measured in:

1. **Same rest-frame wavelength** (e.g., sdss-u band in the local universe but sdss-i at $z = 0.8$). Are you sampling the same stellar population with your image?
2. **Same physical spatial scale**—this is key. If you can resolve HII regions in one set of images but not in another, do not compare!
3. **Same image signal-to-noise**—see the above comments. Morphometric parameters change with s/n and their associated segmentation choices.

This is to emphasize while one can compare different morphometric catalogs to one another, to do so with a healthy dose of skepticism. Your rest-frame ultraviolet images at $z = 6$ ultra cannot be straightforward be compared to the local universe catalog from say Lotz et al. (2004). **The best way to use morphometrics is within the same survey!**

A heroic effort to compare asymmetry across HI surveys can be found in Reynolds et al. (2020). Different sensitivity and resolution (both angular and physical because of distance) meant that the asymmetry values had to be accounted for both differences in sensitivity and resolution.

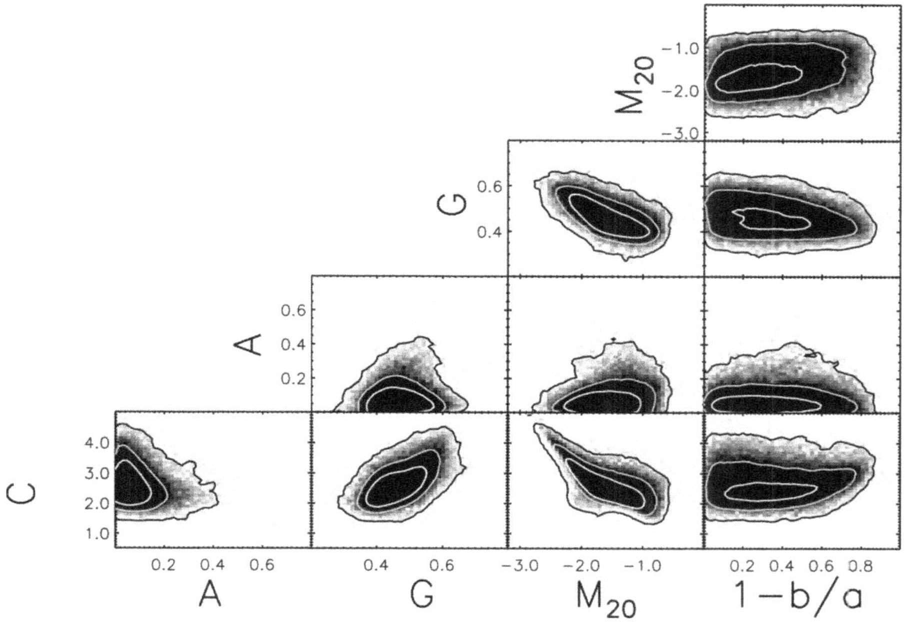

Figure 6.5. Relations between the morphometric parameters (M_{20}, G, A, C, and ellipticity ($1 - b/a$)). Contours enclose 30% (white contour), 80% (gray contour), and 98% (black) of all the COSMOS galaxies. The correlations or degeneracies among some of these parameters, M_{20}, C, and G, are clearly visible. Figure reproduced from Scarlata et al. (2007).

6.9 Orthogonalizing

The morphometric parameter space is far from orthogonal. Gini and concentration are linked, but not the same. M_{20} and Asymmetry probe some but not all of the same features. A logical step is to orthogonalize one's morphometric catalog (Conselice 2006). This is what a Principal Component Analysis (PCA) does. It is equivalent to finding the eigenvectors of a parameter space. With this in mind (Scarlata et al. 2007) used PCA to quantify morphology with ZEST (Zürich Estimate of Spiral Type). The number of parameters going in is illustrated in Figure 6.5: several parameters are degenerate. Once orthogonalized, one is left with three independent components, as illustrated in Figure 6.6. The five morphometric parameters have been reduced to three independent dimensions. Within that new orthogonal space, sub-spaces can then be identified with morphological classes, e.g., edge-on galaxies, bulge dominated galaxies etc.

6.10 Morphometric Uses

6.10.1 Hubble Types

The original goal of galaxy morphometrics was to provide a parameter space in which galaxies could be quickly assigned a Hubble type. This succeeded in so far that one can assign a very broad category of Hubble type to a galaxy. For example

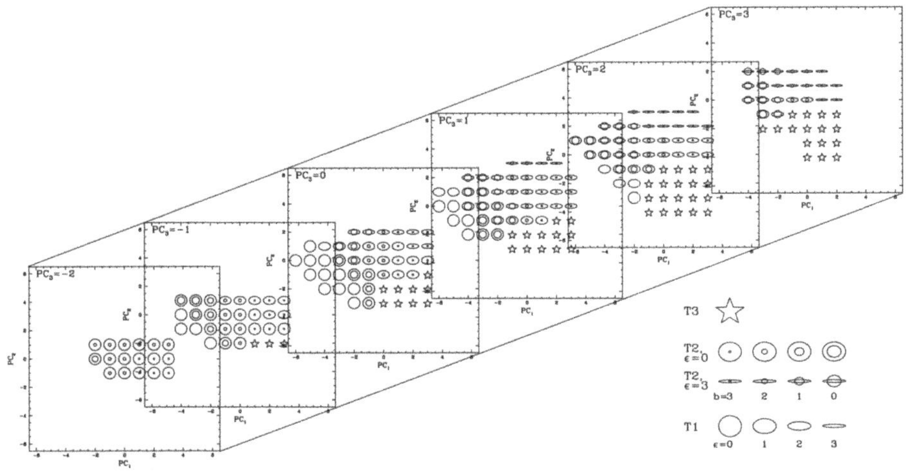

Figure 6.6. ZEST morphological classification. Shown are slices of the principal component (PC) space at constant value of the third component (PC3). Each slice shows PC1 versus PC2; the value of PC3 increases from bottom left to top right. In each unit cube the symbol indicates the ZEST classification associated with that cube (according to the legend shown in the bottom right corner of the figure). In particular, stars represent $T = 3$ galaxies, and single ellipses represent $T = 1$ galaxies. The double ellipses indicate $T = 2$ galaxies, for which the size of the inner ellipse increases for increasing bulgeness parameter. Figure reproduced from Scarlata et al. (2007).

Figure 6.2 shows the Gini–M_{20} parameter space (Lotz et al. 2004) where disk and spheroidals clearly occupy well-separated parts of the diagram. However, there is substantial scatter from one group into others.

6.10.2 Merger Identification

A second use of morphometrics was to identify galaxies that were "disturbed" i.e., undergoing a significant amount of tidal interaction. In Figure 6.7 the part of the parameter space that seems reserved for morphologically disturbed galaxies seems apparent. If one wants to know how important galaxy mergers were at any point in the past 10 Gyr of galaxy evolution, one could generate plots like that for galaxies nearby and far. That should work ok as long as the images over which the morphometrics were taken are of the same physical scale and the wavelength for the nearby and distant (past) sample are the same: use Sloan nearby and Hubble for more distant galaxies. this gives us merger fractions and numbers at different times: say 20% of all galaxies at $z \sim 1$ is merging, aka N per cubic Mega-parsec.

What then remains is an indication how long galaxies stay in the merger part of the diagram to get the merger rate: number of mergers per volume of space over time. This can be explored using simulations of mergers (see also Chapter 10) and one can go into some detail: what is the ratio of galaxy masses? What is the gas fraction of the merging galaxies. Traditionally, 1:1 down to 1:3 merger is considered a "major merger" and anything below 1:3 a "minor merger." This nomenclature is not terribly descriptive so perhaps think of these are morphologically "transformative" and "survivable."

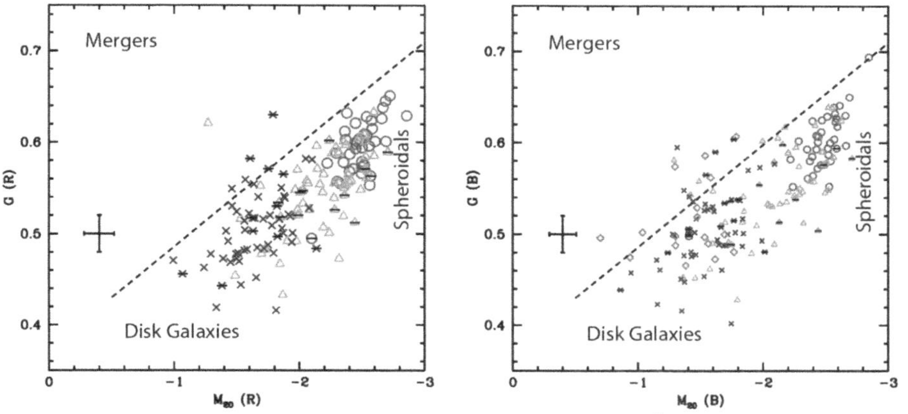

Figure 6.7. The Gini–M_{20} parameter space for R-band and B-band images. Crosses are disk galaxies, triangles S0 and circles ellipticals. A broad classification seems possible in this parameter space. The dashed line denotes the demarcation between normal galaxies and merging ones. Figure adapted from Lotz et al. (2004). © 2004. The American Astronomical Society. All rights reserved.

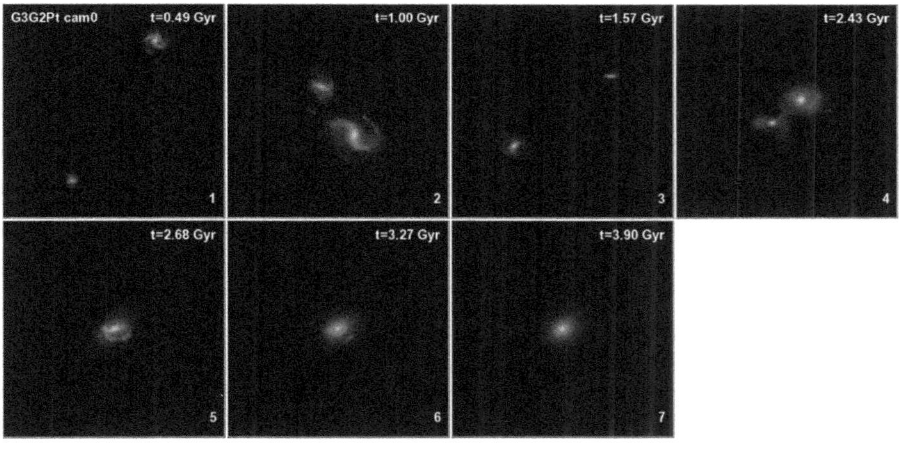

Figure 6.8. One view of two galaxies merging in one simulation of the suite presented in Lotz et al. (2010a). The merging stages are approach, first and second pass, merger, and post-merger. Figure reproduced from Lotz et al. (2010a). © 2010 The Authors. Journal compilation © 2010 RAS.

Figure 6.8 shows one viewing angle on a merger in a simulation (Lotz et al. 2010a, 2010b). One can then measure the morphometrics as both galaxies approach, during first and second close passage, actual merger, and post-merger stages. These are visualized in Figure 6.9 with color codes for the different stages. Note how merging galaxies certainly are morphologically disturbed (high asymmetry, Gini or Concentration values) but certainly not the entire time of the merger. Disturbed morphology is a sign of the first or second close pass and the actual merger. Out of the 4.5 Gyr time the merger took, the occupy a 0.5–1 Gyr in the "merging" part of morphometric parameter space.

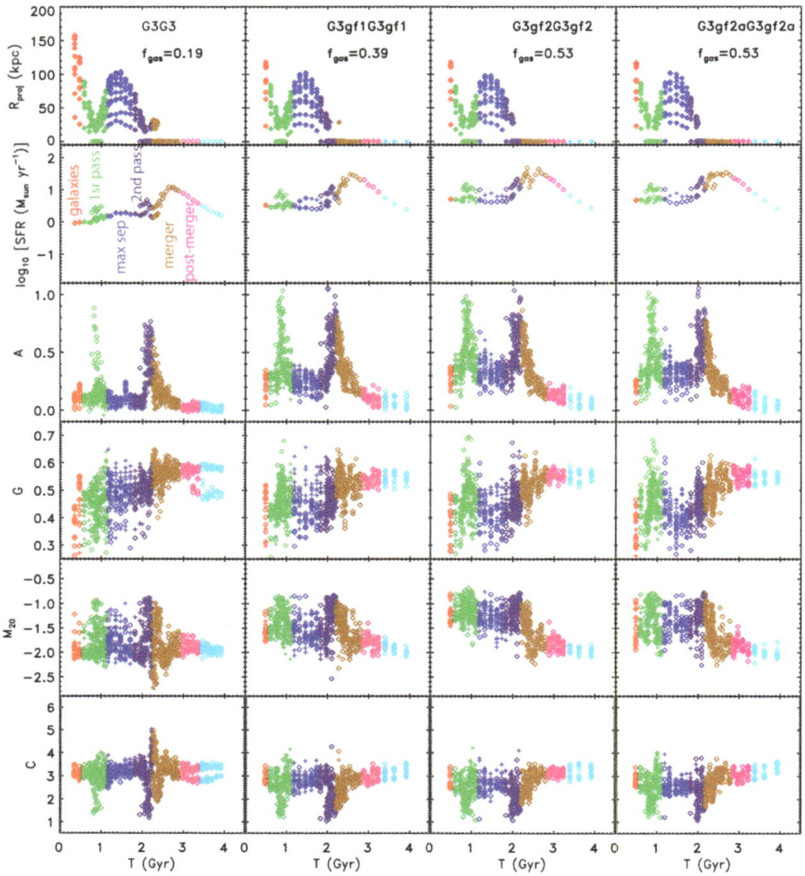

Figure 6.9. The evolution of four different equal-mass merger simulations over a period of 4.5 billion years. Merger stages are galaxies (red), first pass (green), maximum separation (blue), second pass (purple), merger (orange), and post-merger (magenta and cyan). The parameters plotted are projected separation ($R_{\rm proj}$), the star formation rate ($\log_{10}$(SFR)) and four morphometrics (A, Gini, M_{20}, and C). Each pair are equal mass but with different gas fractions ($f_{\rm gas}$ = 19, 39 and 53% gas). The last column is a higher resolution re-run of the third column simulation. Figure adapted from Lotz et al. (2010a). © 2010 The Authors. Journal compilation © 2010 RAS.

The level of disturbance also depends on how "major" the merger actually is. Figure 6.10 shows the same measurements but for two new sets of merger simulations, one set with 1:3 mass ratio and one with a 1:9 mass ratio. Note how mergers happen quicker when the mass ratio is lower and how galaxies spend less time looking "disturbed." These are all considerations for using morphology metrics as a way to measure merger rate over cosmic times.

The different identifications through morphometrics of mergers themselves remain a little controversial: Figures 6.11 and 6.12 show the morphometrics for nearby galaxies (similar to SDSS galaxies) but observed to much greater depth and higher resolution as part of the Kilo Degree Survey (KiDS; de Jong et al. 2019; Kuijken et al. 2019). Figure 6.11 shows the Gini–M_{20} space and where one expects

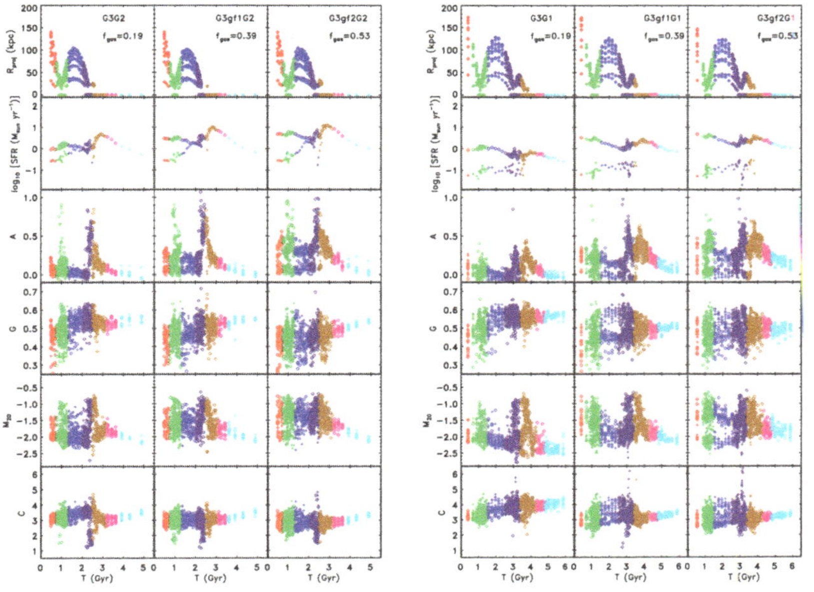

Figure 6.10. The same as Figure 6.9 but for a 1:3 and a 1:9 mass ratio merger. Mass ratio decreases overall merger time and how long it is visible as abnormal morphometric values. Figures reproduced from Lotz et al. (2010a). © 2010 The Authors. Journal compilation © 2010 RAS.

mergers to lie. Figure 6.12 shows where they lie in the Asymmetry–Smoothness metric. It looks like the A–S metric agrees better with identifications done visually (Darg et al. 2010a, 2010b). And this is true *in this particular instance of survey data!* Smoothness has a tuning parameter that make it optimal for a particular resolution and there is a transition visible in Figure 6.11. From the simulations, we already know that a merger spends some time in "normal" morphometric space. Hence better to conclude that Asymmetry agrees best with visual identification of mergers.

Cross-Checks

If we have learned anything from galaxy morphometrics is that these parameters can be an informative space, potentially easing classification but at the same time, they vary with image depth and resolution, observed wavelength, redshift (changing rest-frame wavelength), and often instrument (noise characteristics and PSF structure). It makes these parameterizations non-trivial[7] to inter-compare between surveys or authors thanks to different implementations. This is a lesson we learn time and again: classifications work within a survey and scheme but they break as soon as two surveys, implementations or human classifiers are compared.

[7] Science speak for frikkin hard.

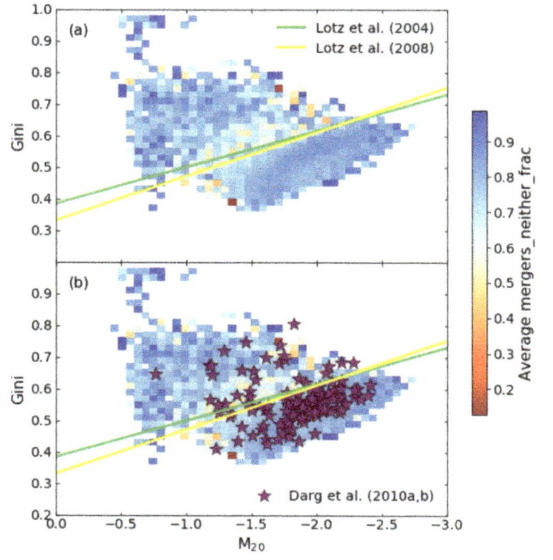

Figure 6.11. Gini versus M_{20} for galaxies in GAMA/KiDS binned by Gini and M_{20}. The voting fraction in favor of *no merger* by the Galaxy Zoo project is shown as the color from low (red) to high (blue). The green and yellow lines are the Lotz (Lotz et al. 2004, 2008) criteria to differentiate between merging and non-merging galaxies. In the bottom panel the galaxies visually identified as mergers (Darg et al. 2010b, 2010a) in KiDS survey images are shown as purple stars. Figure reprinted with permission from Pearson et al. (2019) © ESO.

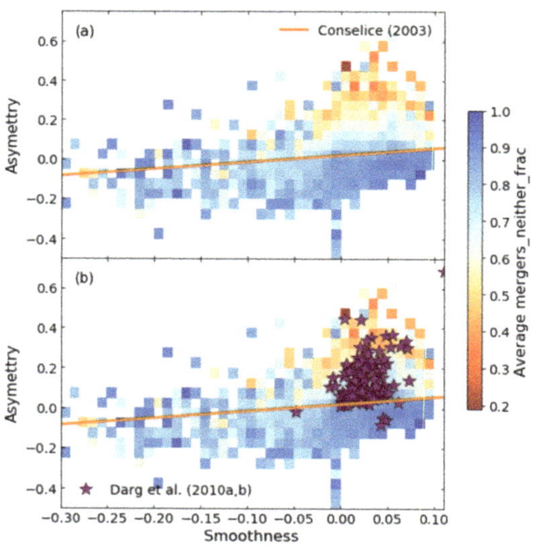

Figure 6.12. Asymmetry (A) versus Smoothness (S) for all GAMA–KiDS galaxies binned by A and S. The voting fraction in favor of *no merger* by the Galaxy Zoo project is shown as the color from low (red) to high (blue). Low scoring areas are where merging galaxies reside. The orange line denotes the criterion for identifying merging and non-merging galaxies from Conselice (2003). In the bottom panel the galaxies visually identified as mergers (Darg et al. 2010b, 2010a) in KiDS survey images are shown as purple stars. Figure reprinted with permission from Pearson et al. (2019) © ESO.

6.11 Assignment

With this section goes the "Morphometrics" assignment (jupyter notebook 10) to illustrate the values of some of the morphometrics on different fake galaxies to build some intuition.

6.12 Further Reading

The two papers that started the widespread use of these parameters are Conselice (2003) and Lotz et al. (2004). Their implementation in STATMORPH (Rodriguez-Gomez et al. 2019) is a good overview of (most) current morphometrics.

References

Abraham, R. G., Valdes, F., Yee, H. K. C., & van den Bergh, S. 1994, ApJ, 432, 75
Abraham, R. G., van den Bergh, S., & Nair, P. 2003, ApJ, 588, 218
Bershady, M. A., Jangren, A., & Conselice, C. J. 2000, AJ, 119, 2645
Bertin, E., & Arnouts, S. 1996, A&AS, 117, 393
Bignone, L. A., Tissera, P. B., Sillero, E., et al. 2017, MNRAS, 465, 1106
Conselice, C. J. 2003, ApJS, 147, 1
Conselice, C. J. 2006, MNRAS, 373, 1389
Conselice, C. J., Gallagher, J. S., Calzetti, D., Homeier, N., & Kinney, A. 2000, AJ, 119, 79
Darg, D. W., Kaviraj, S., Lintott, C. J., et al. 2010a, MNRAS, 401, 1552
Darg, D. W., Kaviraj, S., Lintott, C. J., et al. 2010b, MNRAS, 401, 1043
Davenport, J. R. A. 2015, I really want to find an astronomical application for morphometrics, https://twitter.com/jradavenport/status/571064841344917504
de Jong, R. S., Agertz, O., Berbel, A. A., et al. 2019, Msngr, 175, 3
Fetherolf, T., & Reddy, N.Mosdef Collaboration 2020a, AAS Meeting 235, 145.05
Fetherolf, T., Reddy, N. A., Shapley, A. E., et al. 2020b, MNRAS, 498, 5009
Florian, M. K., Li, N., & Gladders, M. D. 2016, ApJ, 832, 168
Freeman, P. E., Izbicki, R., Lee, A. B., et al. 2013, MNRAS, 434, 282
Giese, N., van der Hulst, T., Serra, P., & Oosterloo, T. 2016, MNRAS, 461, 1656
Gini, C. 1912, Variabilità e Mutuabilità. Contributo allo Studio delle Distribuzioni e delle Relazioni Statistiche (Bologna: C. Cuppini)
Holwerda, B. W. 2005, arXiv:astro-ph/0512139
Holwerda, B. W., Munoz-Mateos, J. C., Comeron, S., et al. 2014, ApJ, 781, 12
Holwerda, B. W., Pirzkal, N., Cox, T. J., et al. 2011a, MNRAS, 416, 2426
Holwerda, B. W., Pirzkal, N., de Blok, W. J. G., et al. 2011b, MNRAS, 416, 2415
Holwerda, B. W., Pirzkal, N., de Blok, W. J. G., et al. 2011c, MNRAS, 416, 2437
Holwerda, B. W., Pirzkal, N., de Blok, W. J. G., et al. 2011d, MNRAS, 416, 2401
Holwerda, B. W., Pirzkal, N., de Blok, W. J. G., & van Driel, W. 2011e, MNRAS, 416, 2447
Kuijken, K., Heymans, C., Dvornik, A., et al. 2019, A&A, 625, A2
Lisker, T. 2008, ApJS, 179, 319
Lotz, J. M., Jonsson, P., Cox, T. J., et al. 2011, ApJ, 742, 103
Lotz, J. M., Jonsson, P., Cox, T. J., & Primack, J. R. 2008, MNRAS, 391, 1137
Lotz, J. M., Jonsson, P., Cox, T. J., & Primack, J. R. 2010a, MNRAS, 404, 590
Lotz, J. M., Jonsson, P., Cox, T. J., & Primack, J. R. 2010b, MNRAS, 404, 575
Lotz, J. M., Madau, P., Giavalisco, M., Primack, J., & Ferguson, H. C. 2006, ApJ, 636, 592

Lotz, J. M., Primack, J., & Madau, P. 2004, AJ, 128, 163
Malin, D. F. 1978, Natur, 276, 591
Pearson, W. J., Wang, L., Alpaslan, M., et al. 2019, A&A, 631, A51
Peth, M. A., Lotz, J. M., Freeman, P. E., et al. 2016, MNRAS, 458, 963
Reynolds, T. N., Westmeier, T., Staveley-Smith, L., Chauhan, G., & Lagos, C. D. P. 2020, MNRAS, 493, 5089
Rodriguez-Gomez, V., Snyder, G. F., Lotz, J. M., et al. 2019, MNRAS, 483, 4140
Scarlata, C., Carollo, C. M., Lilly, S., et al. 2007, ApJS, 172, 406
Takamiya, M. 1999, ApJS, 122, 109
Yitzhaki, S. 1991, American Statistical Association, 9, 235

Chapter 7

Low Surface Brightness

The low surface brightness regime ($\mu > 27$ mag arcsec^{-2}) holds clues to events in a galaxy's history. In a way they are scars of violent events long ago or subtle hints of recent growth. Low surface brightness pertains to the outskirts of galaxies mostly as features made of a relatively small number of stars, spread over many tens to hundreds of kpc will wash out near the main body of a galaxy.

7.1 Observing LSB

Many of these tidal features were already uncovered in the age of large-scale photography. The wide-view glass plates in Schmidt type telescopes are ideal to pick up faint features surrounding galaxies. So while many of such features were already known, much less could be said about them. Detector artifacts or reflections in the telescope can easily pose as low surface brightness (LSB) features. There two ways we can observationally discern LSB features: integrated light and counting resolved stars in deep images. Both are tricky and fraught with technical pitfalls.

7.1.1 Integrated Light

Low surface brightness features were first found using integrated light and photographic plates. It helps that these are sensitive, wide-view, and could be integrated for long times. With modern equipment, these streams stand out in very deep imaging with integration times best measured in hours or days. Small telescopes, with a wide field-of-view can capture the galaxy itself and the faint wispy stream next to it.

The first systematic search for low surface brightness features around galaxies beyond the Local Group has been made by Martínez-Delgado et al. (2010), Martinez-Delgado et al. (2015), with a survey report in Martinez-Delgado (2019).

doi:10.1088/2514-3433/ac2c7dch7

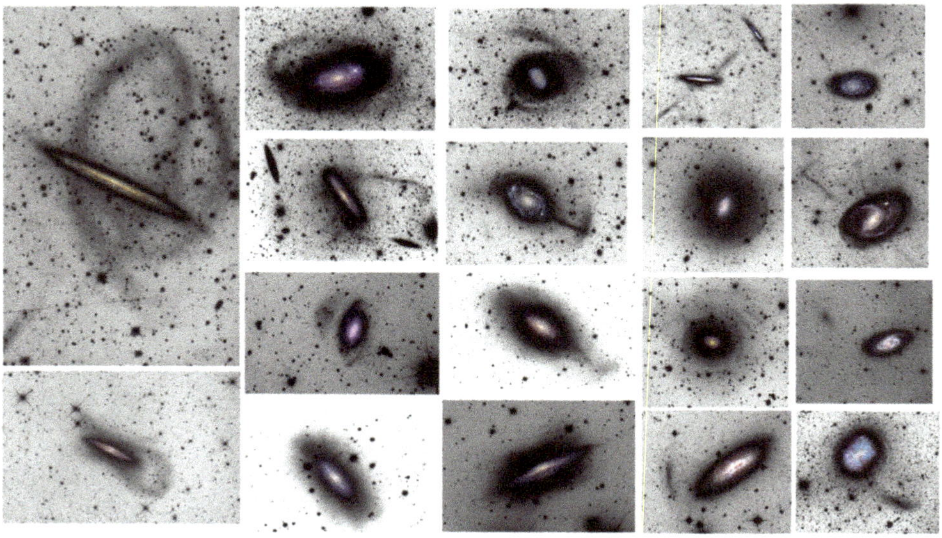

Figure 7.1. Luminance "white" filter images of nearby galaxies from the Stellar Tidal Stream Survey (Martinez-Delgado 2019) showing large, diffuse light structures in their outskirts. A color inset of the disk of each galaxy has been overplotted for reference. Figure received with permission from Martinez-Delgado (2019).

This group uses small "amateur"[1] telescopes for tens of hours to produce deep images such as those in Figure 7.1. A great review of this amateur/professional combined effort can be found in Martinez-Delgado (2020).

These first streams were discovered around galaxies where there was anecdotal (e.g., via photographic plates) but some more structured approaches (with dedicated programs on small telescopes) have also been done, e.g., The Halos and Environments of Nearby Galaxies (HERON) Survey (Rich et al. 2016, 2017; Mosenkov et al. 2020a). But by far the most successful integrated light observational program is the Dragonly Telescope (Abraham & van Dokkum 2014; van Dokkum et al. 2014) which has produced results on a host of LSB phenomena (Abraham et al. 2016; Merritt et al. 2016a, 2016b; Danieli et al. 2017, 2018; Zhang et al. 2018; van Dokkum et al. 2019b; Gilhuly et al. 2020).

These integrated light observations can point us to how often LSB phenomena happen and what their shapes are. However, many observational phenomena can mimic of confuse the detection of low surface brightness features: there is light reflected of dust in our Milky Way (cirrus) and structure in the material in our own solar system (zodiacal light) as well as a host of telescope reflections and artifacts. Figure 7.12 illustrates how a well-known and established feature can remain controversial.

[1] A persistent misnomer, "amateur" astronomers are not paid but volunteer their time, equipment, and expertize and often produce data on par or better with professionals. "Backyard" astronomer does not work either, these setups are in prime dark locations for optimal sky viewing.

7.1.2 Integrated Light Challenges

There are a host of subtle issues that can compromise LSB observations (Knapen & Trujillo 2017), in no particular order:

1. Sky Brightness
2. Galactic Cirrus & Zodiacal Light
3. Internal Reflections & Scattered Light
4. Flat-Fielding & Drift Scanning
5. Masking and Background Subtraction.

Sky brightness is the light from the night's sky—airglow—supplemented with reflected light from the Moon. Depending on the wavelength, this sets a threshold of received light, ideally without a gradient.

Zodiacal light is light reflected or emitted by the dust in the ecliptic plane of the solar system. This is all the small grains left by planetary bodies.[2] It is fairly smooth (goes over degrees) so it shows often as a gradient in a deep image.

Galactic Cirrus is reflection/emission from the dust in interstellar matter dust clouds at higher Galactic Latitude i.e., not the dust lane in the plane of the Milky Way but some scattered clouds above and below it. How much of a problem these are is very dependent on where one is observing. Figure 7.2 shows two fantastic examples of structures cirrus. It is not just diffuse or smooth and reflections of cirrus can pose a problem. In some fields, it will be hard to find a "dark" patch of sky (Figure 7.3).

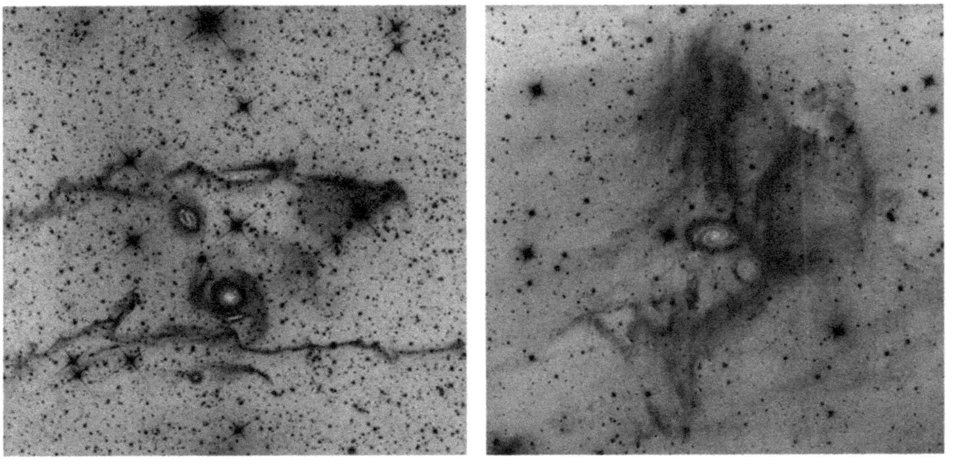

Figure 7.2. Galactic cirrus shows in deep images (10 and 5 hours exposures with 0.5 m telescope) of two nearby galaxies NGC 2634 (left) and NGC 918 (right). These are clouds of material belonging to our own Milky Way and easily mimic complex structure in surrounding a nearby galaxy. Figure reprinted with permission from Martinez-Delgado (2020). © 2019 International Astronomical Union.

[2] Everything from volcanic ejecta from Jupiter's moon to Elon Musk's Tesla car. It all reflects light.

Figure 7.3. Galactic cirrus show in glorious color by the Dragonfly telescope. There is some color-dependence for cirrus. Between the cirrus and the foreground stars, it can be tricky to find a dark spot. Image reprinted with permission from the Dragonfly Telephoto Array, Copyright © 2019 (https://www.dragonflytelescope.org/gallery.html).

Flat-Fielding & Drift Scanning are two techniques related to CCD image processing. If the image is a "stare-into-space" the slight irregularities in the CCD become apparent and one needs to correct for the difference in pixel-to-pixel efficiency (some pixels in a CCD are 99.9% efficient, others just 98%, never an issue when the topic is bright but now it is). So one needs to take a long exposure with uniform illumination (stare at a blank wall with camera) and correct for that. Do that incorrectly (smudge on wall) and that can start mimicking LSB structure. An alternate way to use a CCD camera on the sky is to "drift-scan," reading out the instrument while the sky drifts past (align telescope/camera with Earth's rotation). Now the variation in pixel row efficiency is not only position dependent but time dependent (but it is the average of all the pixels in each row, so more consistent that way).

Reflections & Scattered Light is what you want in a telescope but some of the surfaces are not supposed to reflect light (support structures) and yet they do. A bright source will therefore result in light elsewhere in the field, sometimes very far away from the original bright source. And even in the detector itself, light scatters away from bright sources (the wings of the PSF) and ends up well away from where one would say that source is. Figure 7.4 shows how bright stars in the field contribute scattered light well away from their position.

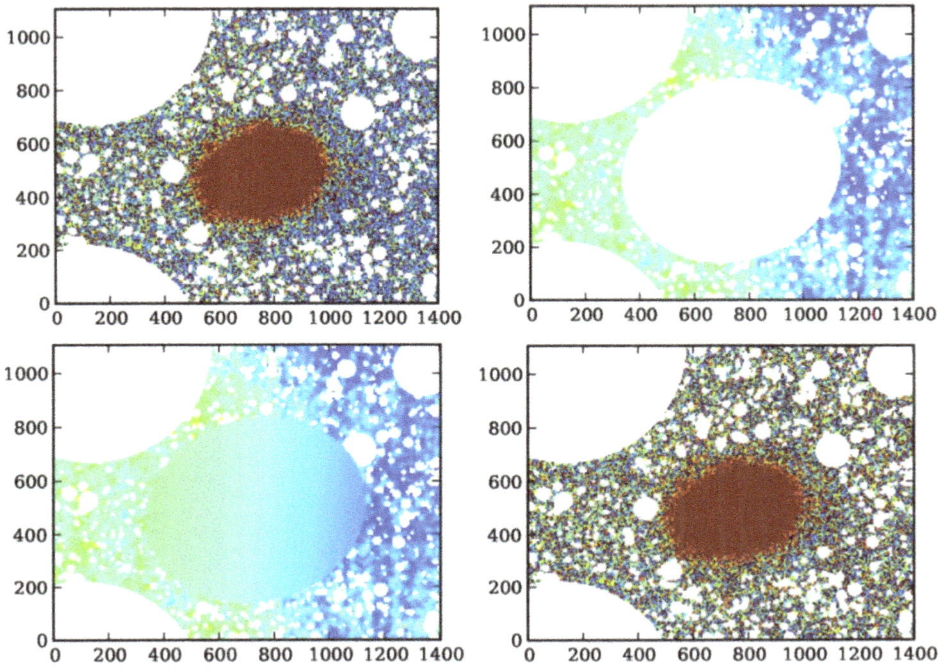

Figure 7.4. Background subtraction example (NGC 941, SDSS Stripe 82). Top left panel: original image, showing a gradient in the background level. White areas are masked-out images of foreground stars and background galaxies. Top right: model for the background, excluding the area of the galaxy. Lower left: background model, extrapolated over the area of the galaxy. Lower right: final image, with the background model subtracted from the original. Gradients in the background light will only become evident after all the bright sources are masked and without correction will mimic LSB features. Original figure from Peters et al. (2017), reprinted with permission from Springer Nature: from Knapen & Trujillo (2017). Copyright © 2017, Springer International Publishing AG.

Masking & Background Subtraction is what needs to be done to a deep image to get rid of all the light that cannot be contributed to the source of interest. One needs to mask the myriad of stars that belong to our Milky Way and the many distant galaxies that will be visible. Once that is done, the subtle gradients in light that remain will become visible and will need to be accounted for. Figure 7.4 illustrates how after generous masking of all the other distinct sources, there is a gradient left in the image.

The short of all these issues is that once your science goal is another magnitude deeper in surface brightness, each of these will need to be accounted for. Thought you masked everything out? Think again. Thought the flat-fielding was perfect. Now it shows structure at the new surface brightness level. This is one example of astronomical observation where as soon as the next level of accuracy is required, one needs to go and improve each of these sources of bias and noise (Figure 7.5).

7.2 LSB from Star Counts

The second observational method to identify LSB features is to observe individual stars (often only the brightest) and count them in a given area of sky. This can only

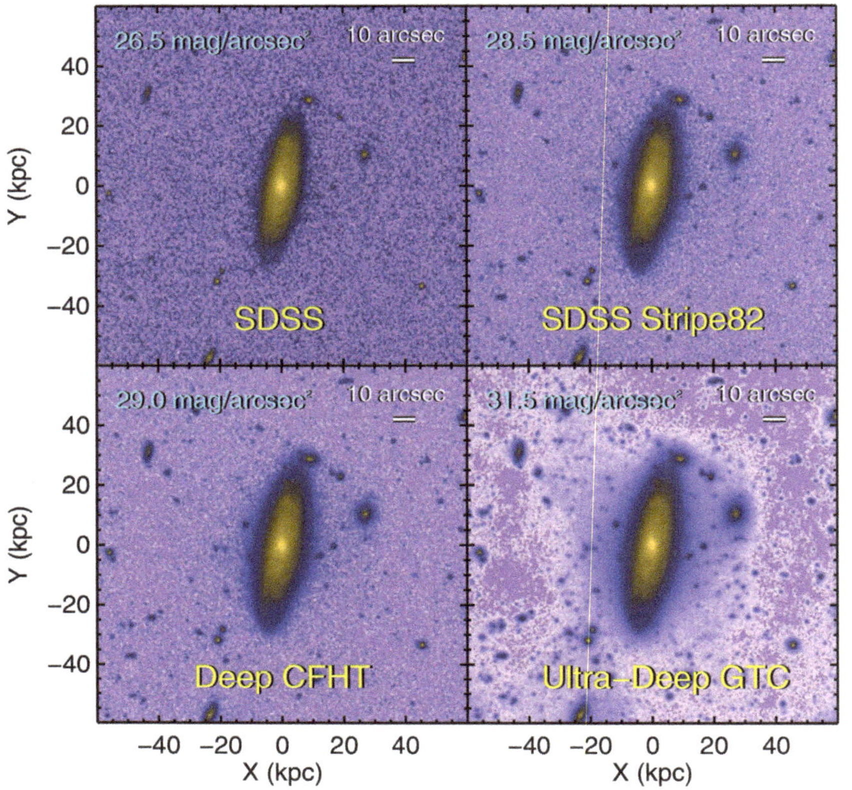

Figure 7.5. UGC 00180 as it would be observed by different surveys: SDSS, SDSS Stripe82, Deep data with Canada–France–Hawaii Telescope (CFHT), and the Gran Telescopio Canarias (GTC). SDSS images are drift scanning imaging and CFHT and GTC are deep stares. Figure reproduced with permission from Trujillo (2016).

be done if one can reliably can get rid of contamination such as dwarf stars in our Milky Way and small, much more distant galaxies. As a result, this method works best in nearby galaxies and only with the highest quality images, i.e., most often Hubble Space Telescope images.

An absolute favorite to count stars around is the Andromeda galaxy (M31) since in its case one can discriminate stars from contaminants even in ground-based images (Ferguson et al. 2002). The Pan-Andromeda Archaeological Survey (PAndAS; McConnachie et al. 2009; Martin et al. 2009; Mackey et al. 2010; Huxor et al. 2014) has been singularly successful in identifying low surface structures around Andromeda (Figure 7.6), identifying 13 different structures (McConnachie et al. 2018). The mix of an external view and a relatively complete census of LSB structures has made PAndAS image the gold standard to compare against integrated light results (Figure 7.7; bottom right).

Just because we are in the middle of it, does make it possible to observe LSB features in our own Milky Way. Famously the Sagittarius stream stands out once the right stellar sub-population is isolated (Ibata et al. 2001; Majewski et al. 2003; Belokurov et al. 2006). With the advent of all-sky and deep imaging and astrometric

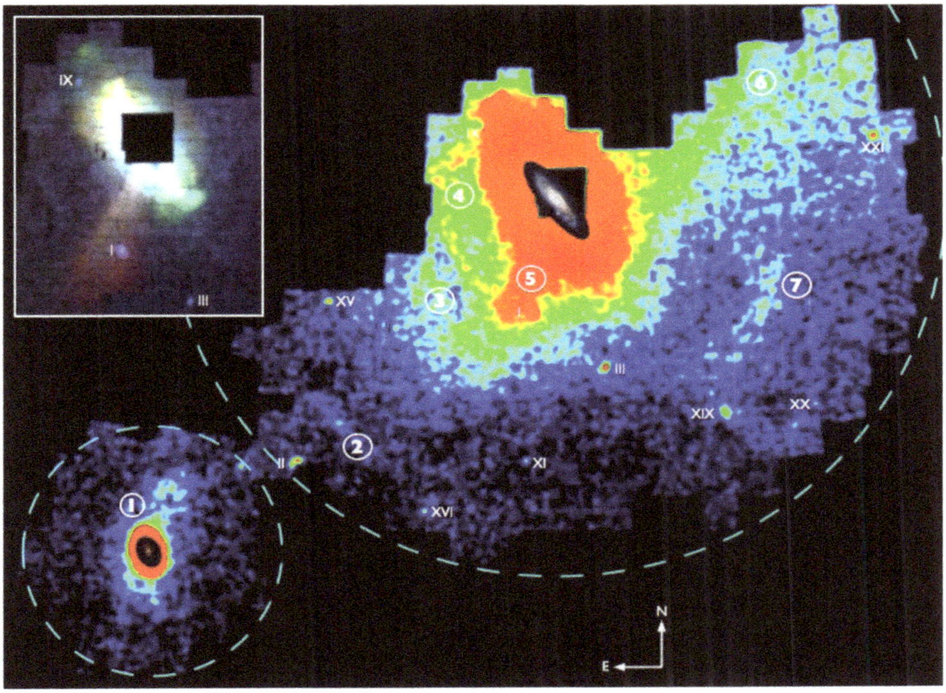

Figure 7.6. Stellar density map of Andromeda—Triangulum. A tangent-plane projection of the density distribution of stellar sources in the PAndAS imaging is shown, with colors and magnitudes consistent with RGB stars at the distance of M31. The inset shows the central parts of the survey at higher resolution. Dashed circles represent the maximum projected radii of 150 and 50 kpc from M31 and M33, respectively. Scale visible images of the disks of M31 and M33 are overlaid. Visible dwarf satellites are indicated with roman numerals. Numbers in circles indicate the largest and most obvious substructures detected: 1, M33 structure; 2, 125 kpc stream (stream A); 3, stream C; 4, eastern arc (stream D); 5, giant stellar stream; 6, northwest minor-axis stream; 7, southwest cloud. Candidate RGB stars are selected by their position in a color–magnitude diagram for a 12 Gyr population at the distance of M31, using only stellar sources with i-mag < 23.5. The inset was created by assigning red–green–blue to metallicities of $-0.4 < [Fe/H] < +0.2$ dex, $-1.3 < [Fe/H] < -0.4$ dex and $-2.3 < [Fe/H] < -1.3$ dex, respectively. Figure reprinted with permission from McConnachie et al. (2009). Copyright © 2009, Macmillan Publishers Limited. All rights reserved.

measurements, such fossils of mergers will be identifiable in the population of stars of the Milky Way (Helmi et al. 2018; Helmi 2020). Identifying features using star counts can be done with some nearby galaxies, e.g., NGC253 (Bailin & Harris 2009) and NGC 891 (Mouhcine et al. 2010; see Figure 7.13).

The identification of LSB features is challenging but possible in other galaxies however. The GHOSTS survey (Radburn-Smith et al. 2011) has attempted this in other more distant galaxies and exclusively with the Hubble Space Telescope. LSB features such as streams are very difficult to identify in Hubble images since their typical width is quickly that of one of Hubble's cameras. This is where one can expect EUCLID and Roman Space Telescopes to gain much new science.

Hubble sampling around nearby galaxies can reveal much more about them than just images of the light, it reveals details of the stellar populations *in* the faint feature

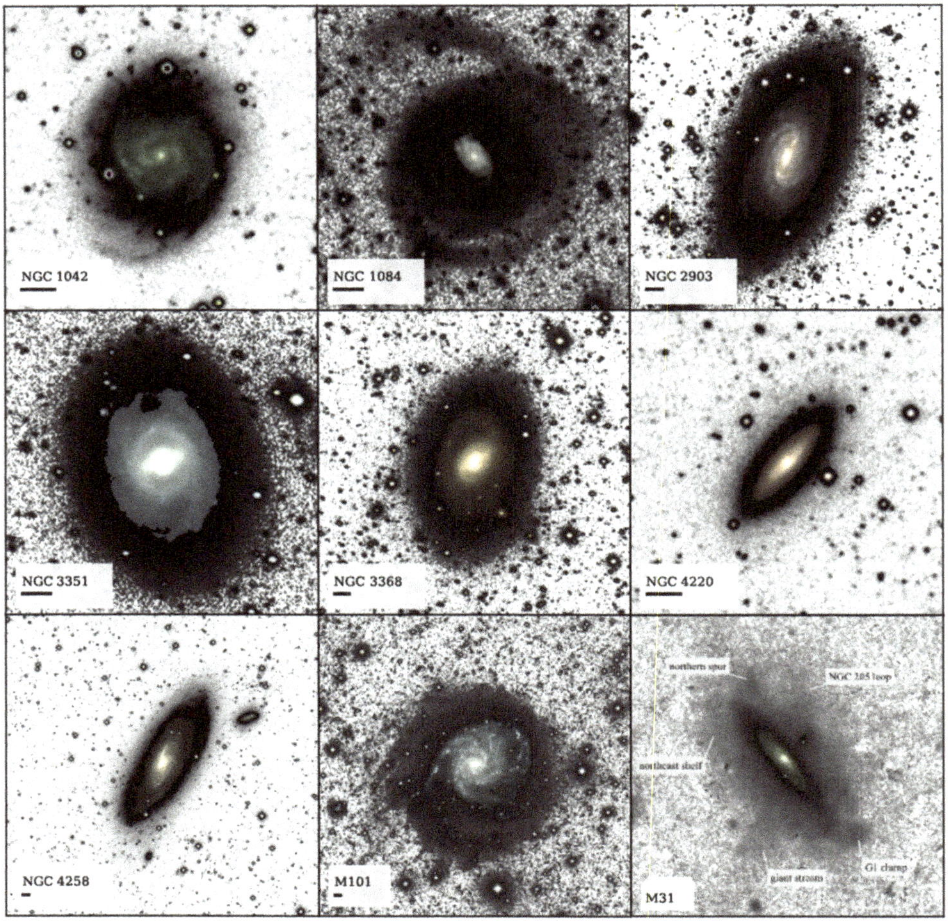

Figure 7.7. Dragonfly images of each of the eight galaxies from Merritt et al. (2016a): pseudo-color images (e.g., SDSS) for the high surface brightness regions, and grayscale for the lower surface brightness outskirts. For comparison the bottom right panel shows M31, created (McConnachie et al. 2009) moved to 7 Mpc (see Figure 7.6). Scale bar is 1′. Similar LSB observations can be made for every galaxy in the New General Catalog (NGC). The profiles (Figure 7.8) and structure of stellar haloes can with a large enough sample be compared to ΛCDM model predictions, constraining our understanding of the timing and pathways for galaxy formation. Figure reproduced from Merritt et al. (2016a). Bottom right figure reproduced from van Dokkum et al. (2014).

such as color and position on the color–magnitude diagram (CMD; Radburn-Smith et al. 2011; Monachesi et al. 2013, 2016). By using characteristics of parts of the stellar population, one can infer metallicity and mean age of the stream (Figure 7.9). Figure 7.10 shows the stellar haloes around the edge-on galaxy NGC 891[3] as mapped by GHOSTS (Streich et al. 2021, in preparation).

[3] NGC 891 is incredibly popular since it is well-known and well-studied, so it attracts comparisons, which attract follow-up observations. Such snowball effects are common in astronomy.

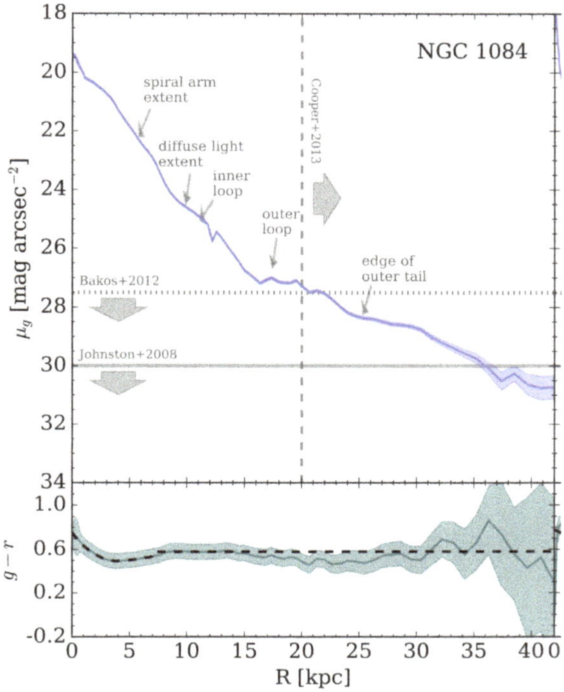

Figure 7.8. An example surface brightness profiles with labeled morphological features as a function of distance along the semimajor axis. Solid horizontal lines highlight the requisite depth to detect faint streams (Johnston et al. 2008), and dotted horizontal lines indicate the surface brightness limit below which M31-like halos are expected to dominate (Bakos & Trujillo 2012). The dashed vertical gray line shows the radius beyond which stellar halos are expected to dominate the light profile (Cooper et al. 2013). The features in the profile of NGC 1084 beyond 7 kpc correspond to the stellar halo. Bottom panels: Observed g–r color profiles. Figure reproduced from Merritt et al. (2016a).

7.2.1 Tidal Features

Tidal features around galaxies are coherent structures that are the result of the tidal disruption of another galaxy after it has interacted violently with the main galaxy. Tidal features is a catch-all and many of the terms used are to describe the different appearance of these features due to slightly different orbits, stage of the interaction, or viewing angle. They break down in broad categories: streams, umbrella's and shells.

7.2.2 Streams

Tidal streams are elongated loops of stars around another galaxy. In some perspectives these can be confused with outer spiral arms and these can have similar size and surface brightnesses in the outskirts (Figures 7.15 A and B, 7.1 and 7.13).

The structural properties (length, width, number of apparent windings around the galaxy) of tidal streams can be used to infer information about the dark matter haloes in which the host galaxies are embedded (Bullock & Johnston 2005; Font et al. 2006, 2008; Johnston et al. 2008). However, most of the tidal stream features

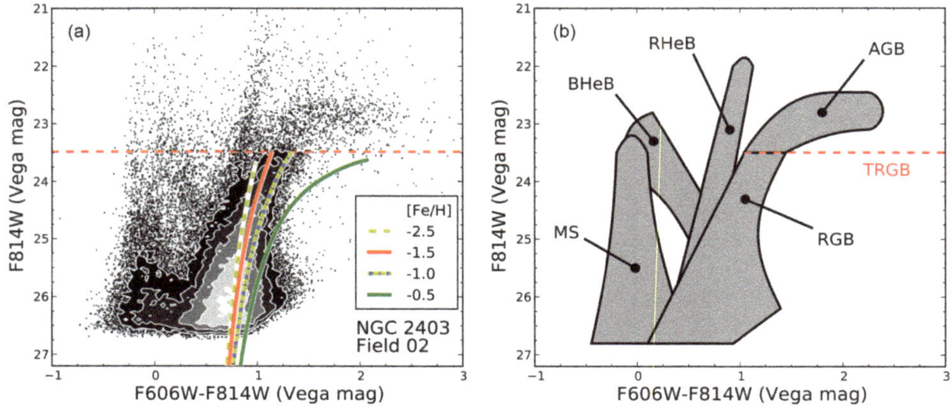

Figure 7.9. How to identify stars that trace the mass? Left is a typical color–magnitude diagram from Hubble two-color images near NGC 2403, which reaches three magnitudes below the tip of the red giant branch (TRGB, dashed red line), is presented in panel (a). Panel (b) identifies the key features of the CMD. These are the main sequence (MS), red giant branch (RGB), asymptotic giant branch (AGB), and red and blue helium-burning sequences (RHeB and BHeB). Also overlaid on panel (a) are a series of lines denoting the color distribution of single age populations of stars (age = 10 Gyr, approximately corresponding to the RGB) for different metallicities taken from the Dartmouth Stellar Evolution Database (Dotter et al. 2008). In this case, the metallicity of the RGB sequence is approximately [Fe/H] = -1.2. Figure reproduced from Radburn-Smith et al. (2011). © 2011. The American Astronomical Society. All rights reserved.

are very low surface brightness ($\mu \gg 28$ mag arcsec^{-2}). While cosmological simulations of Milky Way size galaxies predict tens to hundreds of tidal features, only one is predicted to be observable with current equipment to stand out with enough contrast from the stellar halo (e.g., 30+ mag arcsec^{-2}, see Figure 7.11; Bullock & Johnston 2005; Johnston et al. 2008; Cooper et al. 2010).

More and more streams are being discovered using the integrated light technique (Martinez-Delgado 2020; Martinez-Delgado et al. 2021a; Figure 7.1). A great example of what can be done even with a small telescope can be seen in Figure 7.14. Individual stream discoveries can remain controversial. For example, the stream around NGC 5907 (Figure 7.12) reported by Martínez-Delgado et al. (2009) was not fully reproduced by the Dragonfly observations in van Dokkum et al. (2019b). One of the reported loops was missing! That alters what we infer from that stream (age, progenitor mass etc) substantially. With more and more of these streams being cataloged, one can expect quantitative, rather than qualitative results to constrain how we understand small galaxy mergers.

The amazing part of the information that is hiding in stellar streams. For example, the number of loops is indicative of how old the loop is (one full loop means the satellite galaxy fell in 2 Gyr ago; Johnston et al. 2008) and even in individual cases, a simulation can often be made to match the appearance, brightness and position of these faint streams, building confidence on how we model galaxies and their gravitational potential (see Chapter 8). More and more streams of every stripe are now being discovered from ever deeper imaging surveys (Martinez-Delgado et al. 2021a).

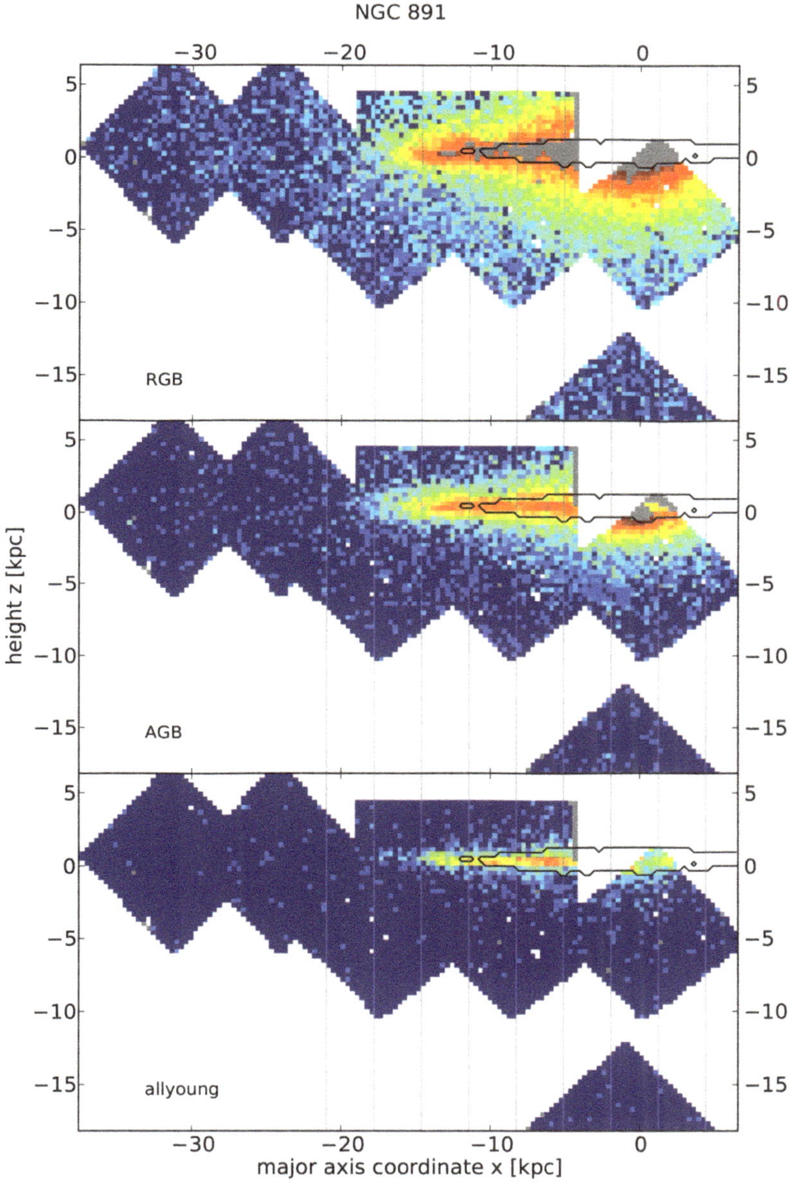

Figure 7.10. Red giant branch stellar density (compare to Figure 7.6) for a nearby edge-on massive galaxy NGC 891. Figure reprinted with permission from Streich et al. (2021, in preparation).

7.2.3 Plumes and "Umbrellas"

A special case of a stellar stream is one where the infalling satellite has hit the bull's eye of the center of the main galaxy. In this case, a large fraction of the stars in the infalling satellite have the same initial velocity and direction but gravitational

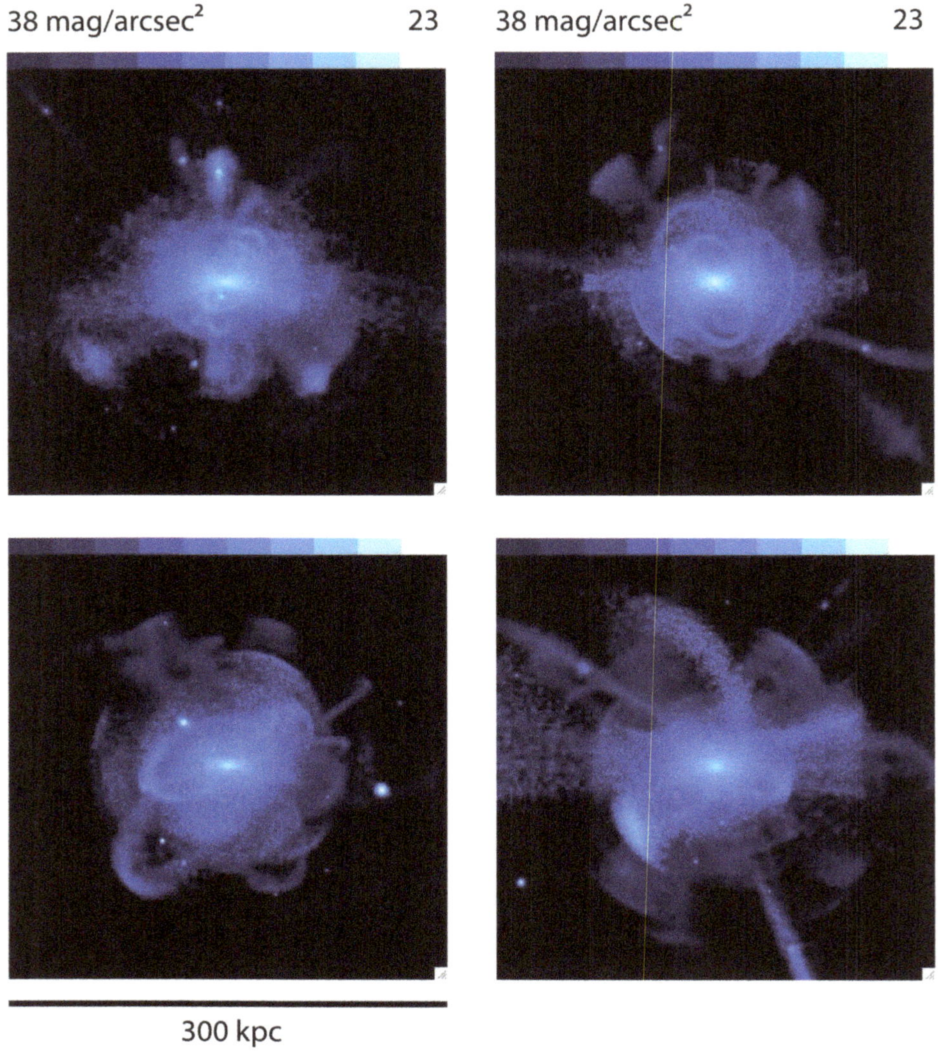

Figure 7.11. The "external galaxy" views for halo realizations from Bullock & Johnston (2005). The boxes are 300 kpc by 300 kpc. The blue/white color scale indicates surface brightness: 23 mag arcsec^{-2} (white) to 38 (dark blue/black). These different realizations of Milky Way haloes illustrate the wide variety of low surface brightness structures and streams expected. Only a very small number of these streams are observable with current equipment. Figure reproduced from Bullock & Johnston (2005).

interaction ("drag") between the satellite's stars and the densest part of the main galaxy cause small changes in the orientations of their orbits: they all had the same kinetic energy coming in and so there is a maximum to their orbit (convert to potential energy) but each has a different major axis of the orbit: their end-point is distributed with respect to the direction of travel.

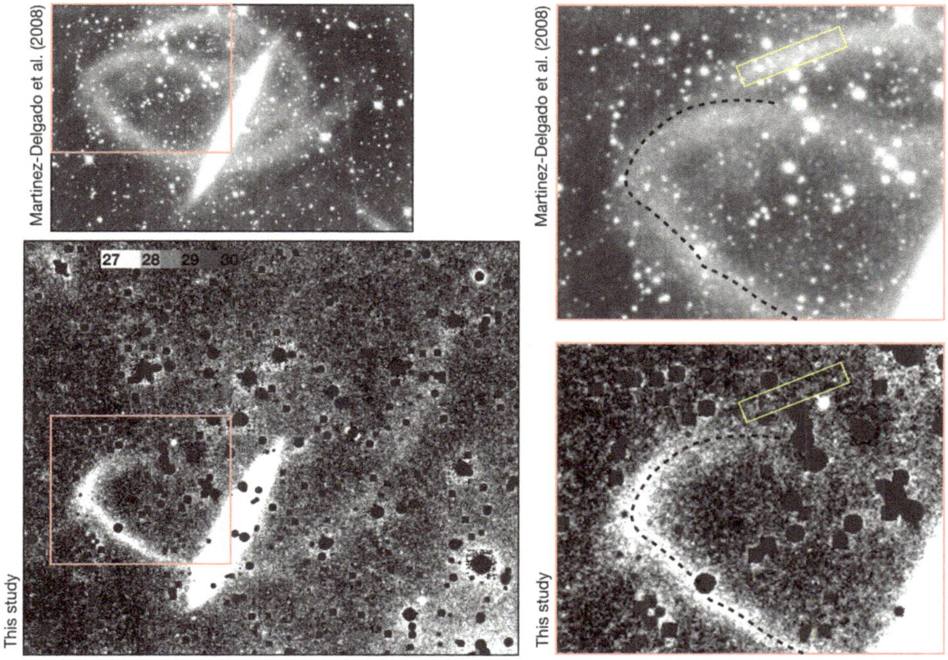

Figure 7.12. The comparison of the stream morphology around NGC5907 by Martínez-Delgado et al. (2009) and in the Dragonfly image by van Dokkum et al. (2019b). The second loop, which is pretty obvious in the top pictures is not reproduced with the other telescope. There is some disagreement where the primary loop is as well. This is to illustrate, that even in the same object, confirmation of such faint features remains extremely challenging. Figure reproduced from van Dokkum et al. (2019b).

7.2.4 Shells

Shell galaxies are galaxies surrounded by ring-like fine structures (cf Figure 7.16). These structures are made of stars and form open, (almost) concentric arcs that do not cross each other. Shells are relatively common in elliptical or S0 galaxies. At least 10% of all elliptical galaxies in the local universe have shells (Malin & Carter 1983; Schweizer et al. 1983; Schweizer & Ford 1985). The number and distances of shells vary, from a single shell to dozens and within 1 kpc to out to 100 kpc (Figures 7.16 and 7.17).

Shells occur when a smaller galaxy merges radially (through the center) of the central elliptical galaxy. As the orbit of the stars is radial, they trade potential energy for kinetic and as they climb back out of the gravitational well of the main galaxy they exchange it back. Small, random motion within this galaxy means that they start close together but scatter perpendicular to the direction of travel. At the apex of these orbits, their velocity is low and these stars cluster at a very specific radius: they form a shell of stars. With each passage, the stars, which pass by many stars that belong to the main galaxy, experience dynamical friction, losing some of their kinetic energy and forming another shell on the opposite side. The shells themselves are faint and are only a few percent of the stars in the galaxy as a whole. These sharp

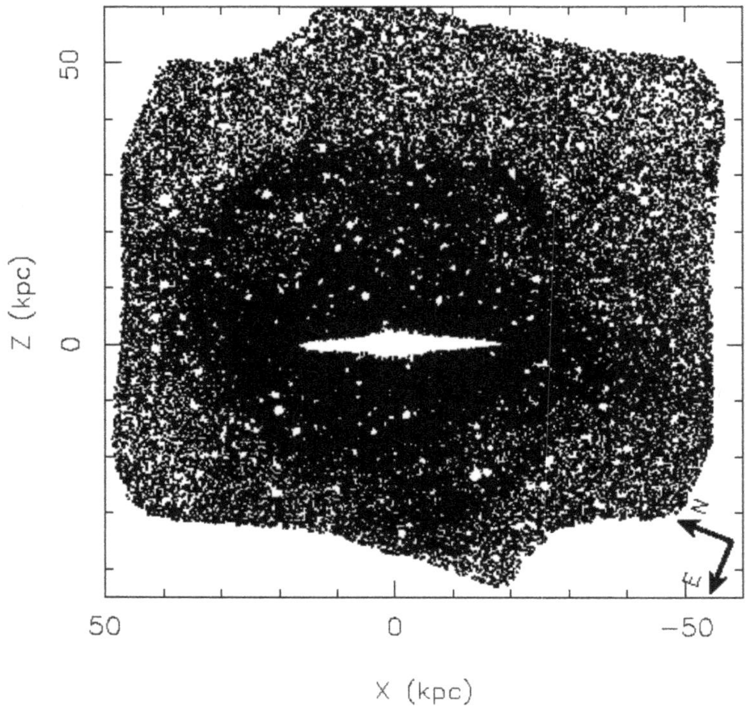

Figure 7.13. The view with Subaru Hyper-Suprime Camera of NGC 891, an edge-on galaxy. By mapping the individual RGB stars, the streams around this galaxy become apparent. Compare to the coverage of the same galaxy by Hubble in Figure 7.10. Figure reproduced from Mouhcine et al. (2010). © 2010. The American Astronomical Society. All rights reserved.

but low-contrast features atop a smooth background of the elliptical galaxy can best be isolated with a technique called "unsharp masking" (Malin 1977). One can think of the "umbrella" or "plume" in the previous section as the first step in forming multiple shells.

7.3 The Stellar Halo

Once multiple smaller galaxies have merged with the main galaxy and mixed their stars, the final result is a spheroidal halo around the main galaxy made from mostly acquired stars. Much like tree rings or sediment, the composition of the stellar halo are indicative of how the main galaxy grew over time. Up to some 10% of all the stars in a massive galaxy can be in its halo! But because they are spread out over a volume many times the main disk's, the stellar halo (with it's structures of streams and plumes etc) barely stands out.

The fraction of stars in the stellar halo is related to the total stellar mass (and dark matter halo mass) of galaxies. Figure 7.18 shows the fraction of stars as a function of stellar mass in galaxies for which reliable measures of the stellar halo exist.

The key prediction by cold dark matter cosmology (ΛCDM) is that massive galaxies are surrounded by a stellar halo made up of the remnants of small, accreted

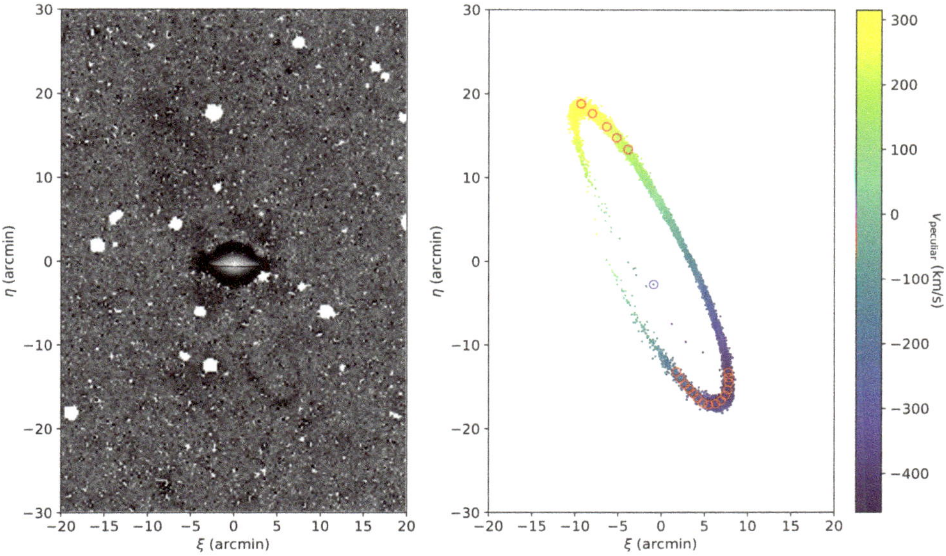

Figure 7.14. M104, the Sombrero galaxy, the nearest spheroidal galaxy. Left panel shows the observed stream around M104 from Martinez-Delgado et al. (2021b). This image has been taken using a small amateur telescope (80 cm) for 4.75 hr total integration. Right panel shows the best-fit stream model colored by the predicted peculiar velocity. This model shows that it is possible to reproduce the observed tidal feature around M104 with a single disruption event. Figure reproduced from Martinez-Delgado et al. (2021b). Copyright © 2021, Oxford University Press.

satellite galaxies (Klypin et al. 1999). Ongoing accretion is evident as faint stellar streams surrounding disk galaxies such as our Milky Way (Bullock & Johnston 2005; Font et al. 2006, 2008). Stars form in the centers of the larger dark matter haloes (White & Rees 1978), where the number of stars that form in low mass haloes is dramatically suppressed compared to larger haloes due to feedback processes. As satellite haloes merge with the main halo, their stellar components tidally disrupt and are spread into a diffuse and structured stellar halo (Bullock et al. 2001; Bullock & Johnston 2005; Cooper et al. 2010; Figure 7.11). The majority of the stellar halo originates from a *single* accretion event (D'Souza & Bell 2018). The resulting stellar haloes are expected to exhibit steep radial density profiles, have abundant substructure (stellar streams) and show considerable halo-to-halo variation in their properties, all of which are expected to be closely tied with their merger histories.

The merger history of galaxies is critically linked to our understanding of ΛCDM structure and galaxy formation. As such, stellar haloes are a strenuous test of the validity of the ΛCDM paradigm. Observational efforts are now underway to detect both stellar streams (Martínez-Delgado et al. 2010, 2012, 2016; Martinez-Delgado 2020) and (Merritt et al. 2014, 2016a) using fast optics telescopes such as the Dragonfly telescope. However, detections remain controversial due to the intricacies of night sky subtraction (van Dokkum et al. 2019b).

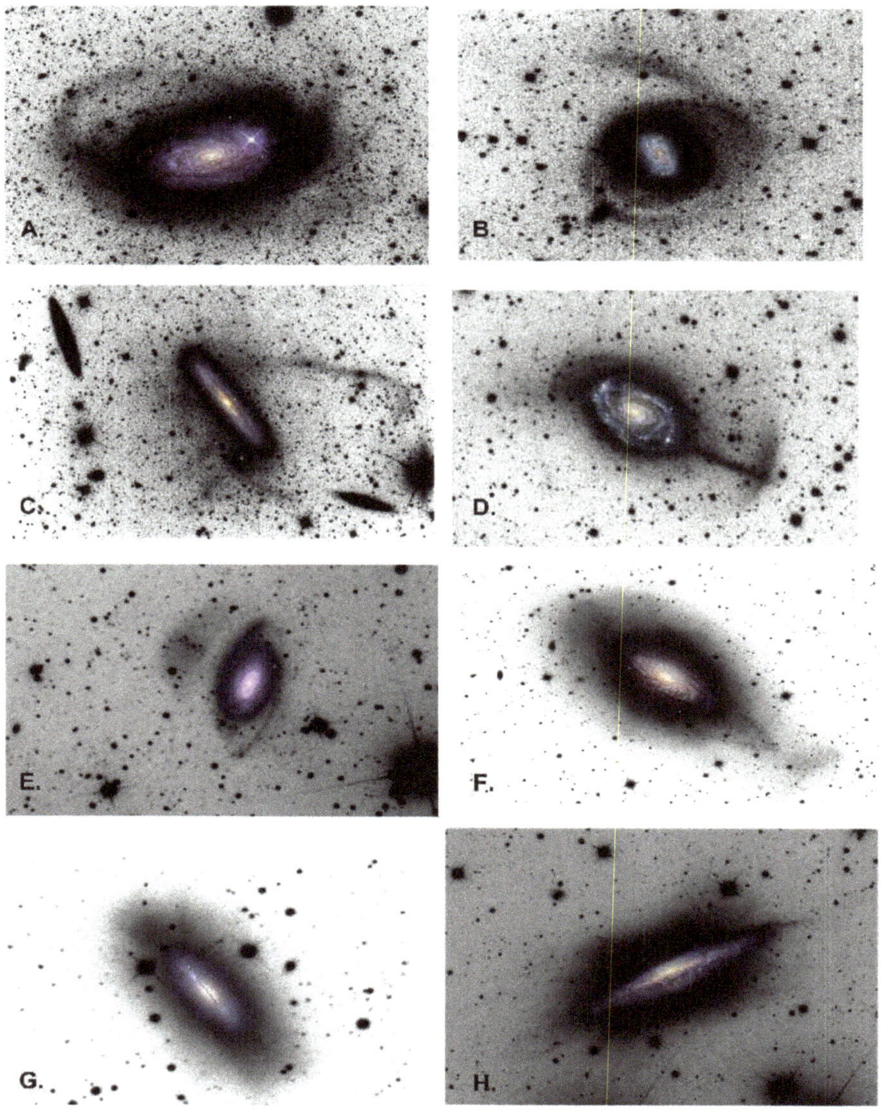

Figure 7.15. Deep images of nearby galaxies showing large, diffuse light substructures in their outskirts: (a) a Sgr-like stream in Messier 63; (b) giant plumes around NGC 1084; (c) partial tidally disrupted satellites in NGC 4216; (d) an umbrella shaped tidal debris structure in NGC 4651; (e) an enormous stellar cloud in NGC 7531; (f) diffuse, large-scale and more coherent features around NGC 3521; (g) a prominent spike and giant wedge-shaped structure seen emanating from NGC 5866 (top right of the galaxy); and (h) a the inner halo in NGC 1055 showing multiple elongated structures. Figure reproduced from Martínez-Delgado et al. (2010).

7.3.1 M101: A Milky Way Analog with No Halo?

With our understanding of galaxy formation predicting that there should be stellar haloes around big disk galaxies, it came as a surprise when it was reported that

Figure 7.16. NGC 474, an example of a shell galaxy. Not only does it show multiple stream-like features but sharp shells stand out as well. Image credit: Jean-Charles Cuillandre (CFHT) & Giovanni Anselmi (Coelum) https://apod.nasa.gov/apod/ap180206.html.

M101, a well-studied face-on galaxy had practically no stellar halo as observed with the Dragonfly telescope (van Dokkum et al. 2014; see Figure 7.19).

This discovery is a great example of the observational science that is astronomy. One technique's discovery can be verified with another. Using Hubble Space Telescope observations of the surroundings of M101 (Figure 7.20), a different team of astronomers could verify the radial profile observed with the Dragonfly and add independent estimates of the stellar halo. Figure 7.21 shows the stars Jang and his team found with Hubble and the densities that translated to. Comparing the Hubble observations to the Dragonfly ones (Figure 7.20). Figure 7.22 extends the disk to these outlying HST fields to estimate how much of the observed stellar population is stellar halo and what fraction is the furthest reaches of the stellar disk.

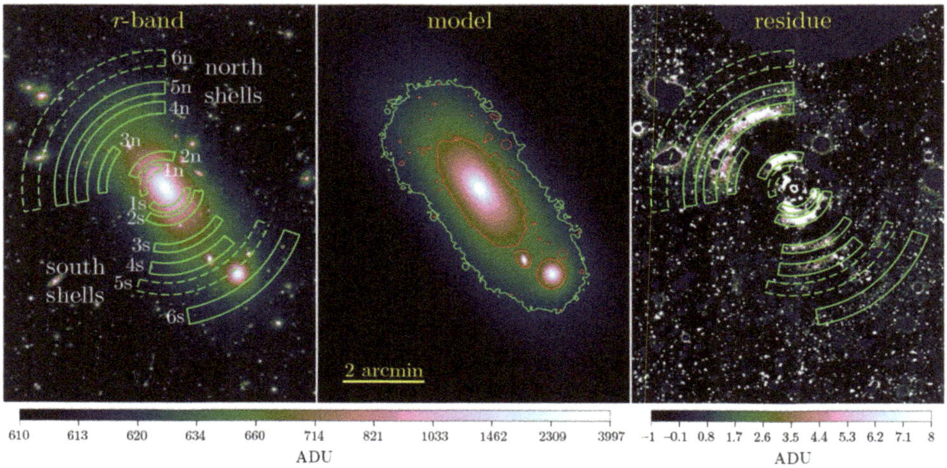

Figure 7.17. Best result of the 2D fit using galfit. Left: original r-band image with the positions of the various shells labeled from 1 to 6, with "s" or "n" referring to the southern and northern shells, respectively. The shells were drawn as circular arcs, symmetric with respect to the center of NGC 4104. Middle: best-fit model (with a flat background included), simultaneously taking into account the three brightest galaxies in the group. The isocontours from the original r-band image, corresponding to 24 and 26 mag arcsec^{-2} (shown in red and green, respectively) are displayed for reference. Right: residue, i.e., original image minus model, with the positions of the shells superimposed. In the left and middle images, the background component is present and the same color scale is used. Figure reprinted with permission from Lima Neto et al. (2020) © ESO.

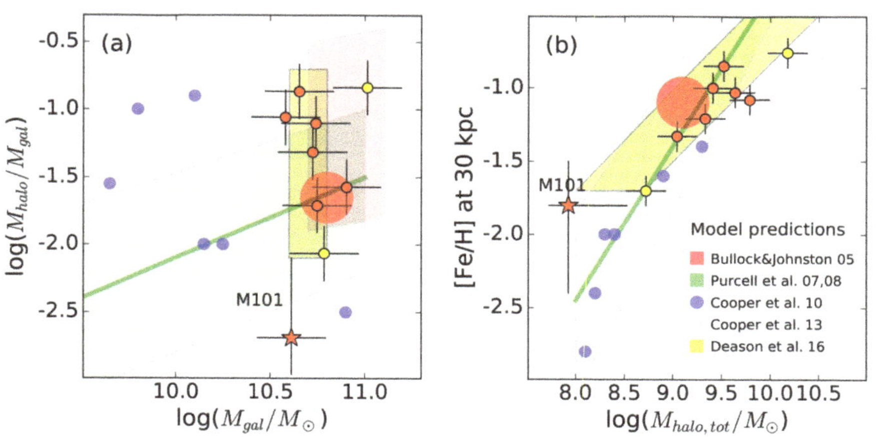

Figure 7.18. Left panel: the stellar mass in the stellar halo as a function of stellar mass of the main galaxy. Right panel: the metallicity of the stellar halo as a function of stellar mass of the main galaxy. Models from Bullock & Johnston (2005), Purcell et al. (2007, 2010), Cooper et al. (2010, 2013), Deason et al. (2016) are the shaded regions and blue points. The observed values from the GHOSTS survey with Hubble (Radburn-Smith et al. 2011; Harmsen et al. 2017; Bell et al. 2017; Streich et al. 2016) are the red dots and M101 (Jang et al. 2020) the red star. There is good broad agreement between models and observations with M101 remaining somewhat outside expected values. Figure reprinted with permission from Jang et al. (2020) © ESO.

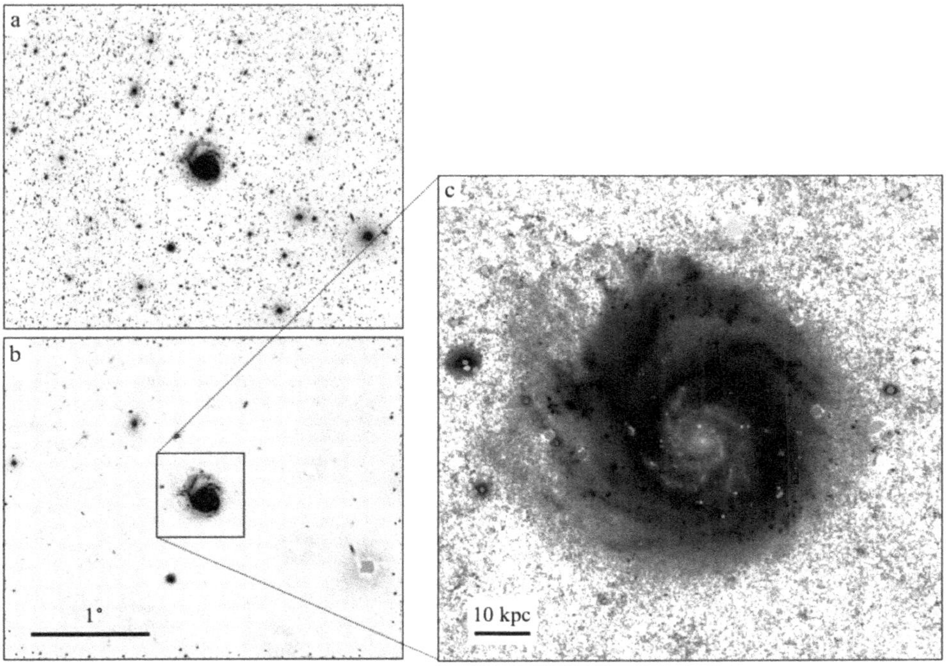

Figure 7.19. The face-on galaxy M101, a favorite of astrophotographers and professional astronomers alike. Panel (a) The Dragonfly telescope g-band image of the area centered on the galaxy M101. North is up and East is to the left. Panel (b) The same image, after subtraction of stars and a model for large-scale background structure. Panel (c) The zoom-in around M101 at high contrast. The faint spiral arms on the East side of M101 (Mihos et al. 2013) have a surface brightness of $\mu_g \sim 29$ mag arcsec^{-2}. The color image was created from the g and r exposures. Figure reproduced from van Dokkum et al. (2014).

The numbers of stars in these HST fields can only be explained by a two component model: a disk and a stellar halo. The Hubble star counts agree with the integrated light Dragonfly result of a very anemic stellar halo. The mass of M101 is right where models of galaxy formation expect a wider range of halo fractions and M101 is at the lowest end of that expected stellar halo mass (Figure 7.18). This one galaxy did not break the models just yet!

7.4 Thick Disks

The thick disk is another low surface brightness feature. The vertical profile of galaxies appeared to consist not of one exponential components but two (Yoachim & Dalcanton 2005). This second disk is at much lower surface brightness but more extended than the thin, high surface brightness one. This is tricky to observe as the Point-Spread Function of the telescope redistributes light from the high surface brightness disk in a way that can mimic a thick disk (de Jong 2008; Comerón et al. 2018).

In the context of ΛCDM galaxy assembly, disks predominantly form through the accretion of gas which is converted into new stars. The spheroidal components of

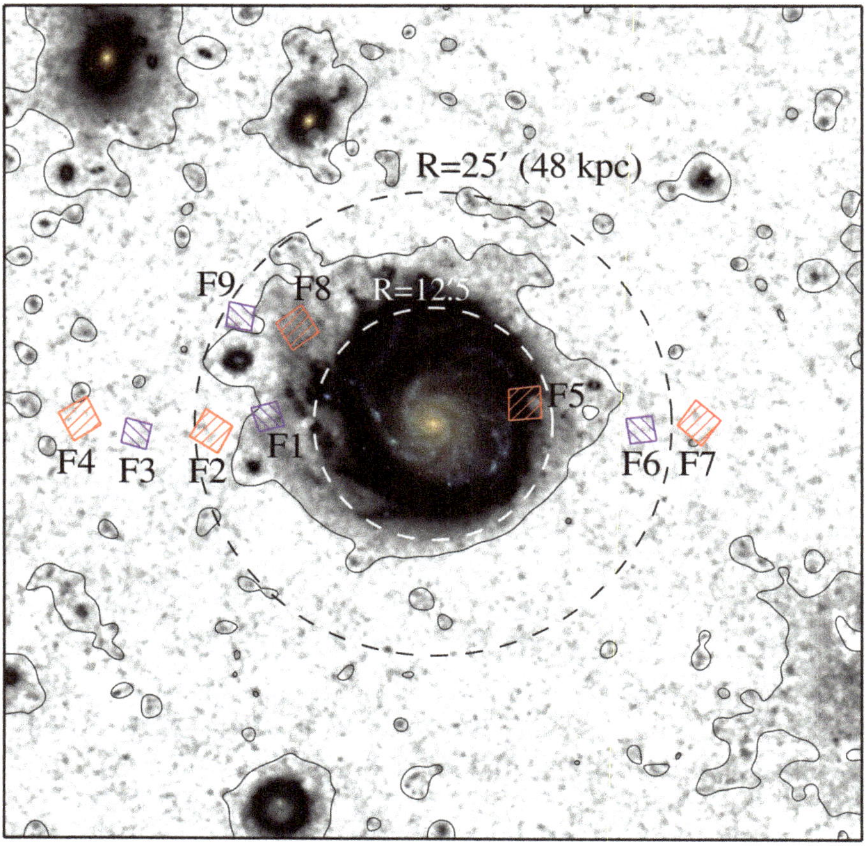

Figure 7.20. The locations of Hubble observations (ACS = red squares and WFC3 = purple squares) overlaid on the deep Dragonfly g-band image from van Dokkum et al. (2014; Figure 7.19). The central high surface brightness regions of image are replaced by the color image from the Sloan Digital Sky Survey. The white and black dashed lines are 24 and 48 kpc from the center of M101. The black contour line is $\mu_g = 30$ mag arcsec^{-2}. Star counts can be sampled for each Hubble observation and compared to the integrated light profile from the Dragonfly. Figure reproduced with permission from Jang et al. (2020) © ESO.

galaxies (bulge and halo) originate mostly from the merger of the stellar systems associated with dark matter subhaloes (Driver et al. 2009). Thick disks are ubiquitous structural components in disk galaxies, discovered first in the Milky Way and then in edge-on external galaxies (Yoachim & Dalcanton 2006; Comerón et al. 2014). They contain important stellar mass fractions, which can be as high as in the thin disk, and give direct clues to the galaxy's formation and evolution

Figure 7.23 shows a series of cross-profiles of an edge-on galaxy. In each there are two components to the vertical light profile. Such thick disks, once seen in the infrared appeared to be ubiquitous (Comerón et al. 2011b, 2011a, 2014, 2015, 2016).

When seen edge-on, some spiral and lenticular galaxies show ample evidence for a disk component beyond the young, thin disk: the thick disk (Yoachim & Dalcanton 2006; Elmegreen et al. 2017; Comerón et al. 2018; Martínez-Lombilla & Knapen

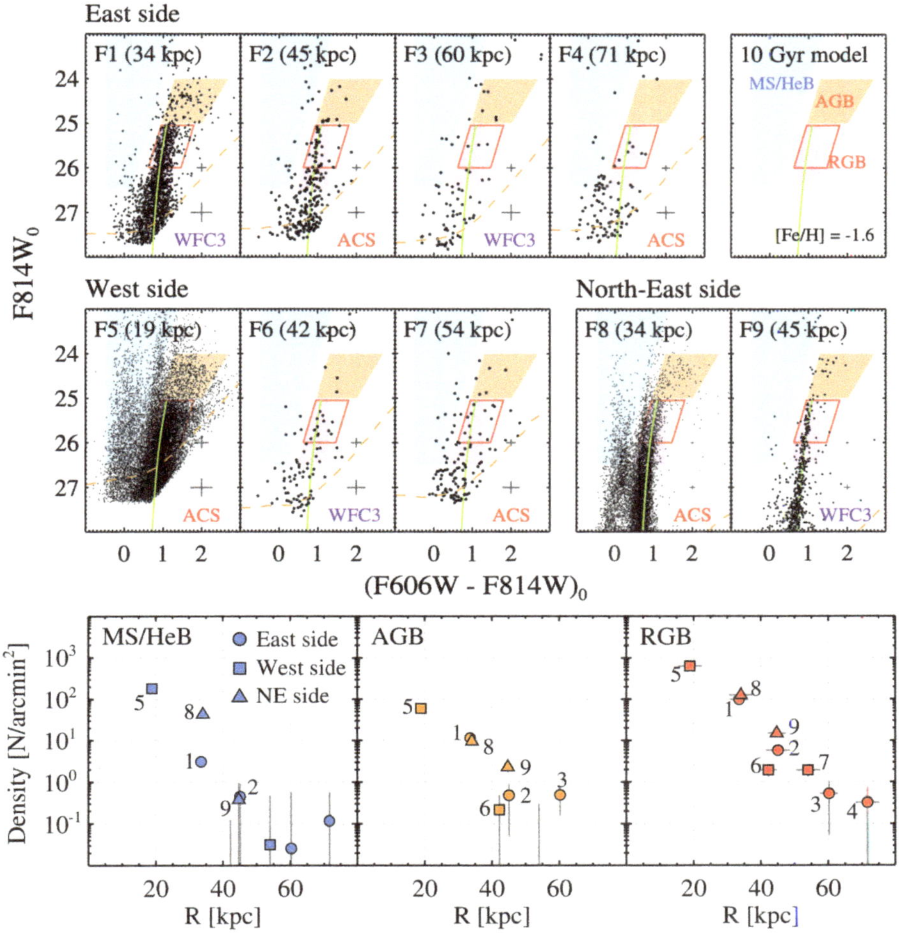

Figure 7.21. The color–magnitude diagram of stars found by Hubble in the fields in Figure 7.20. Like in Figure 7.9, we can identify the brightest main sequence (MS), asymptotic giant branch (AGB) and red giant branch (RGB) stars (model in the top corner). Converting how many we observe in each (and correcting for background effects) the surface densities in the bottom plot are found as a function of radius from the center of M101. Figures reprinted with permission from Jang et al. (2020) © ESO.

2019), as does our Milky Way. The origin of this second component is thought to be either:

(A) **secular:** formed at high redshift and dynamically heated as the thin disk formed subsequently with younger generations of stars, or

(B) **environmental:** induced by a late-stage merger/accretion event (directly or indirectly).

The relative importance of the two scenarios remains unclear. Thick disks may hold an equal stellar mass as the star-forming thin one and the thick disk's longer timescales make it a rare relic of a galaxy's past.

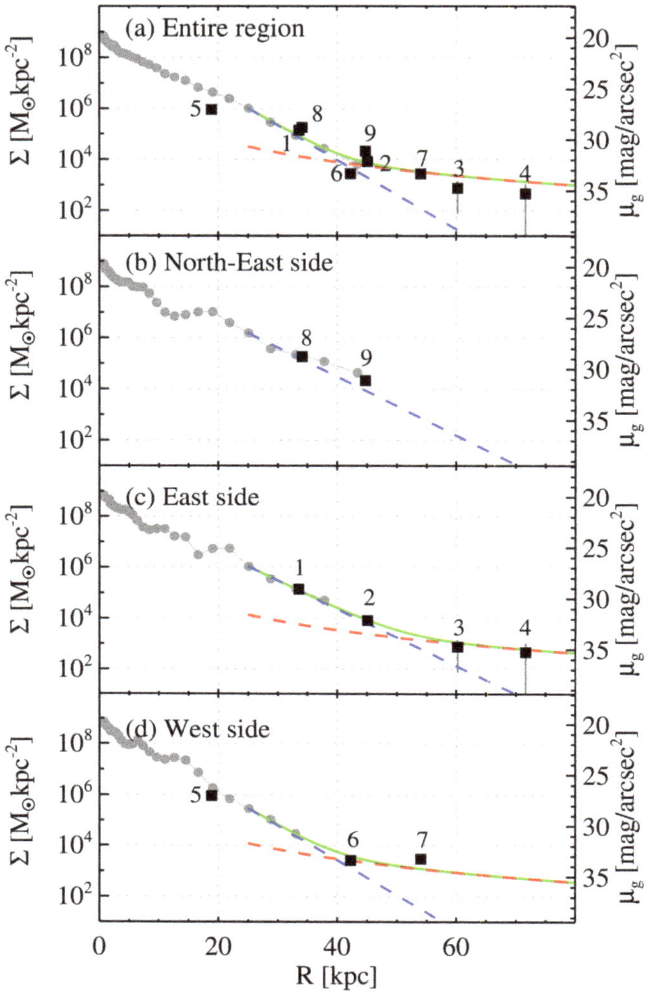

Figure 7.22. Radial mass surface density profiles taken for M101 (a), the North–East side (b), the East side (c), and the West side (d). The gray points are from Dragonfly (Figures 7.19 and 7.20) and the black squares are derived from RGB counts in the Hubble observations (each field labeled in Figure 7.20). The dashed lines are fits for the outer disk (blue) and the stellar halo (red), an exponential and power-law respectively, their sum is the green solid line. The anemic stellar halo found by Dragonfly was confirmed with Hubble RGB counts. Figure reprinted with permission from Jang et al. (2020) © ESO.

In the Milky Way, the thick disk is kinematically and chemically recognizable as a relic of the early formation. The observational picture for our own Milky Way is evolving with both metallicities, ages, and exact positions of disk stars mapped throughout the disk (Bovy et al. 2016). Most of the thick disk's stellar population is considered to have formed *in situ*[4] (Bovy et al. 2012); the structure of the thick disk

[4] Fancy Latin term for "right there".

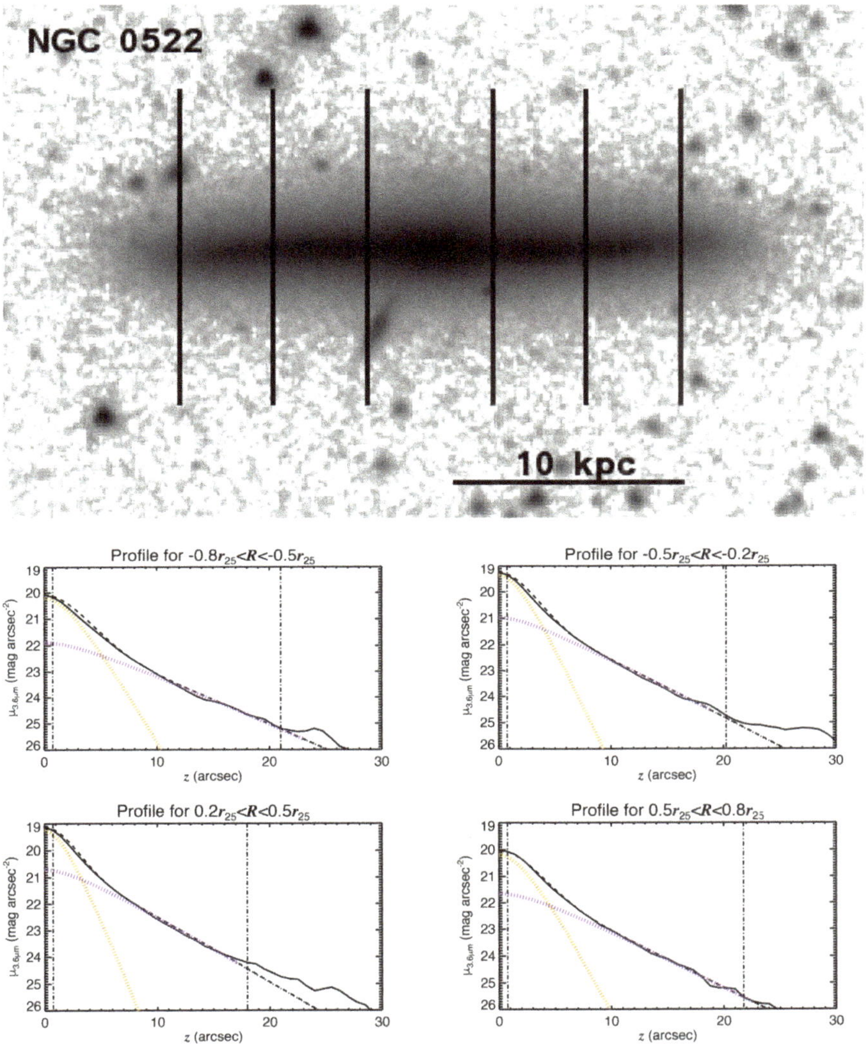

Figure 7.23. Vertical profiles of an edge-on galaxy. Left: the 3.6 μm-band S^{4}G image of NGC 0522. NGC 0522 is similar to the Milky Way both in morphological type and in size. The vertical lines are the parts from which luminosity profiles have been produced, two on the left side and two on the right. Right: the vertical profiles of this galaxy. The solid curve is the observed luminosity profile, and the dashed curve the best fit. The dotted curves indicate the contributions of the thin (orange) and the thick (purple) disk. Figures reproduced from Comerón et al. (2011a).

has been shown by Helmi et al. (2018) to originate mostly from a single mid-sized merger, which brought some 17 kinematically-identified globular clusters (GCs) with it. The good agreement between the properties of that cluster population (Kruijssen et al. 2019) and those of the thick disk (Haywood et al. 2013; Snaith et al. 2014) imply that *a substantial fraction* of the thick disk stellar population was formed at that time (see however Marsakov et al. 2020 for a contrary view).

For other galaxies, the picture remains very mixed. Some studies find very old thick disks (i.e., likely secular: Yoachim & Dalcanton 2006, 2008b; Guérou et al. 2016; Katkov et al. 2019; Martínez-Lombilla & Knapen 2019; Comerón et al. 2015, 2016, 2019; Kasparova et al. 2020) and some find evidence for a recent mass injection into the thick disk, especially in denser environments (environmental: Yoachim & Dalcanton 2008a; Kasparova et al. 2016; Pinna et al. 2019b, 2019a). Both scenarios seem viable ways to build a thick disk, either in external galaxies or own Milky Way. But which scenario is the most common? And is a second age component in the thick disk evidence of a merger?

The role of the environment in formation of thick disks remains therefore a bit controversial. Yoachim & Dalcanton (2008a) point to kinematic evidence for (early) merger origin for many of their thick disks, while Comerón et al. (2019) point out the low odds of only one out of the 17 thick disks with kinematics showing retrograde movement, a unique indicator of a past merger. In contrast, three S0 galaxies of the Fornax cluster do show a component of the thick disk originating from a small merger and part of the thick disk forming early and *in situ* (Pinna et al. 2019a, 2019b). Figure 7.24 shows the maps made with the MUSE Integral Field Unit instrument, a camera that takes a spectrum for each pixel in the image. One can use

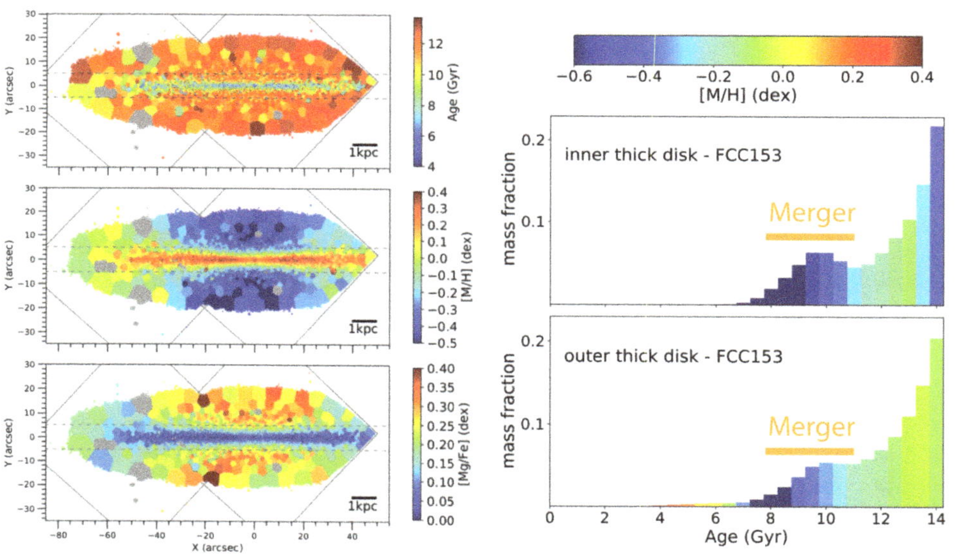

Figure 7.24. Left: stellar population maps for FCC 153. From top to bottom: mean age, total metallicity [M/H], [Mg/Fe] abundance. The discarded bins are plotted in gray, as well as the coverage of the two MUSE pointings. The physical units are indicated by the scale bar on bottom right of each panel. The horizontal dashed lines indicate the thin disk scale height ($\pm z_0$). Right: the inner and outer thick disk stellar age and metallicity distributions for FCC153 (Pinna et al. 2019b) with the merger-induced part of the thick disk marked. The GC ages of a progenitor would need to be >10Gyr as star formation shut off during the merger. Figures adapted from Pinna et al. (2019a) © ESO.

the spectral information to map age and metallicity content of galaxy components using model fits of stellar population templates. Separating out different components, the ages of stars in each can be examined (Pinna et al. 2019b, 2019a). The thick disk components in FCC 153 show evidence for a bump of stars moving into the disk some 10 Gyr ago. With the origin of thick disks now under scrutiny, we can ask ourselves if that was triggered by a small merger or flyby.

Computer simulations similarly are inconclusive: Martig et al. (2014) concludes that a thick disks cannot have a merger origin, it remains unclear if inside-out forming disks should flare (Minchev et al. 2015; Kawata 2018; Beraldo e Silva et al. 2020), but some conclude minor mergers are a viable thick disk formation mechanism (Qu et al. 2011; Martig et al. 2016; Aumer et al. 2016). Yu et al. (2021) make a prediction based on the FIRE-2 simulations for the Milky Way and similar galaxies: some of the TDs in normal environments will be younger if they had a recent merger with an LMC-like galaxy. Their fraction of thick disk due to a minor merger is $\sim$25%.

7.5 Disk Breaks

The edge-on perspective on disks allows us to do a neat trick: observe to lower surface brightnesses in the same exposure time. Because the disk is viewed edge-on the light is "integrated along the line of sight," meaning that the lower surface brightness parts of the disk are summed along our line-of-sight. In the earliest studies of disk profiles (cf van der Kruit & Searle 1981, 1982; van der Kruit 2007), this showed a curious effect: the radial extent of the disk appeared to stop. The disk appears to be truncated. The sudden truncation of the disk is visible above and below the plane as well (Martínez-Lombilla et al. 2019).

Disk truncation is now a well-studied phenomena but the term itself has been replaced with "disk break" because it more accurately describes what happens. The stellar disk does not have a hard limit but there is a clear break in the surface brightness profile; it goes from one exponential over into another. There are three types of disk breaks: I, II, and III.[5]

Each profile is illustrated in Figure 7.25. Type I are single exponential profiles. At no point does the surface brightness profile change from one scalelength to another. Type II is the previously mentioned "truncation" or "downbending"; the disk profile bends down, there are fewer stars in the outer parts with a steeper decline than in inner disk. The scalelength of the inner disk is greater than that of the outer disk ($h_i > h_o$). Type III is "up-bending" or "anti-truncation," where the profile flattens for the outer disk; the scalelength is greater in the outer disk ($h_i < h_o$).

7.5.1 Origin of Disk Breaks

This leaves the question why this visible "edge" to the stellar disk leaves us any clues to the formation or evolution of the galaxy itself. Many models have been developed to explain disk truncation (Type II profile): an angular momentum limit (van der

[5] Sadly astronomy numbers things when they could have been descriptive.

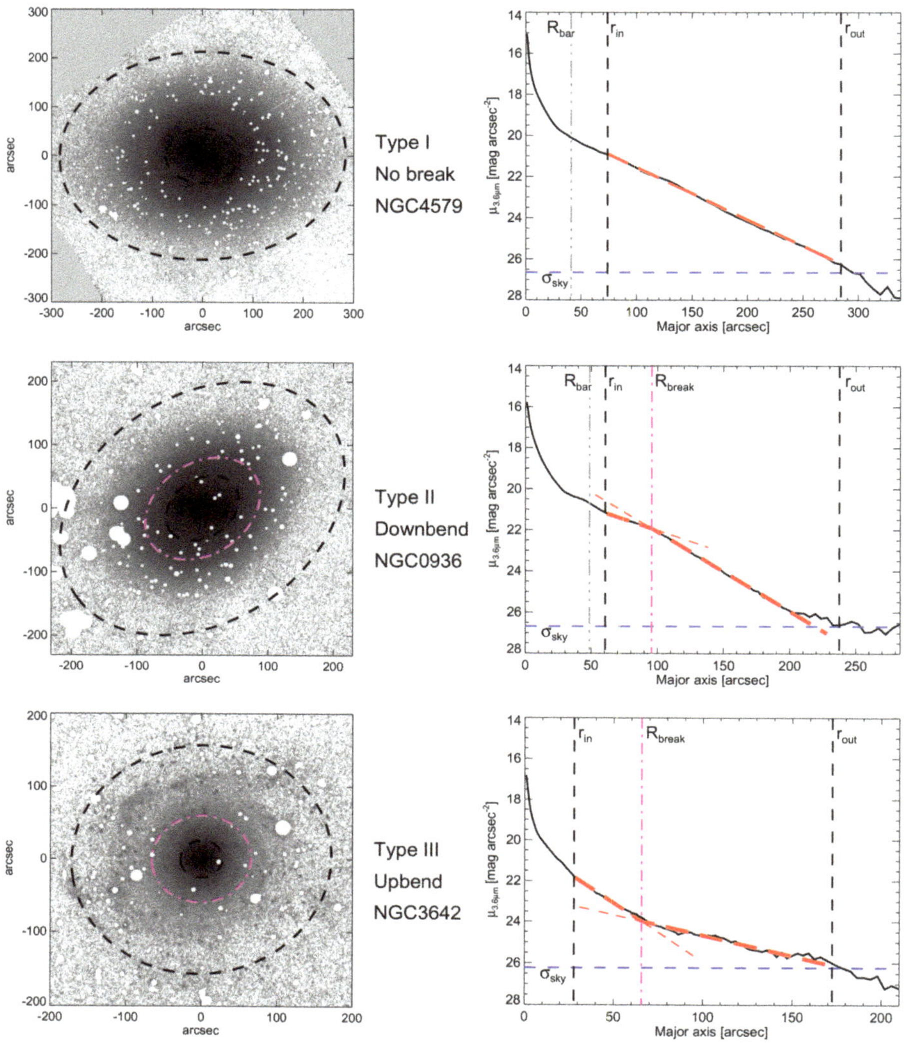

Figure 7.25. Examples of the different disk profile types in the radial surface brightness: Type I, single exponential, Type II, "truncated" or downbending break and Type III, up-bending break. The red dashed line is the fit profile to the radial surface brightness profile for each galaxy. The vertical dashed lines, and the dashed ellipses in the images on left, present the inner- and outer radii of the fit range. The break radius is the dot–dashed line. Figure reproduced from Laine et al. (2014). © 2014 The Authors, Published by Oxford University Press on behalf of the Royal Astronomical Society.

Kruit 1987), a star formation threshold (Kennicutt 1989), disk resonances (Minchev et al. 2012), diffusion by dark matter subhalo interactions (Kazantzidis et al. 2009), or a combination of a star formation threshold and radial migration (Roškar et al. 2012).

Similarly, up-bending profiles (Type III) have many explanations: minor mergers, tidal interactions, bars, and gas accretions (Younger et al. 2007; Kazantzidis et al.

2009; Herpich et al. 2015). The very outskirts of disks form sensitive test areas for these models, because their evolution is more easily affected and extreme due to the low disk self-gravity and the lack of gas for star formation (Figure 7.26).

7.6 Disk Warps

The neutral hydrogen (H I) gas disk of many nearby disk galaxies are warped. Based on a sample of 26 edge-on galaxies, it has been asserted that all HI disks that extend beyond their optical (stellar) disk are warped (García-Ruiz et al. 2002). Unlike stellar warps, the onset of the H I warp is often abrupt, discontinuous, and in edge-on galaxies, it coincides with the break in the exponential disk profile (Briggs 1990; van der Kruit 2007; see previous section). However, the mechanism(s) that form these warps are still unclear (for a review, van der Kruit & Freeman 2011).

One possible explanation is that the warp is a manifestation of ongoing gas accretion (Macciò et al. 2006; Spavone et al. 2010; e.g., the XUV disks discussed in the next section). Many galaxies show possible signs of accretion in their H I outskirts in the form of bridges to companions, asymmetric density or velocity fields, and gas clouds with anomalous velocities (for a review, see Sancisi et al. 2008). Alternatively, tidal effects from neighbors or "sloshing" of the disk within the dark matter halo could be alternate reasons an H I disk is warped.

Figure 7.27 shows NGC 4565, a nearby edge-on galaxy in H I. There are several other H I phenomena going on with this galaxy as well but we focus on the warp on the North side of this edge-on disk. Note that the warp is asymmetric and much more pronounced on the Northern end. It was noted early that disks truncate where H I warps begin, suggesting that the two phenomena have a common origin (van der Kruit 2007). Figure 7.28 shows the stellar populations in the HST image taken at the

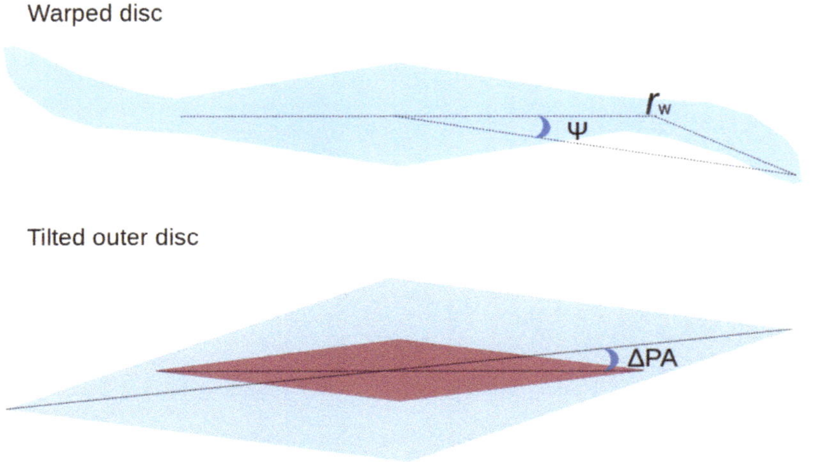

Figure 7.26. The difference between disk tilting and warping. The warp is characterized by an angle Ψ and a warp radius r_W. The angle between the two different planes of the inner and outer disks in the bottom sketch is shown as ΔP_A. Figure reproduced from Mosenkov et al. (2020b). © 2020 The Author(s), Published by Oxford University Press on behalf of the Royal Astronomical Society.

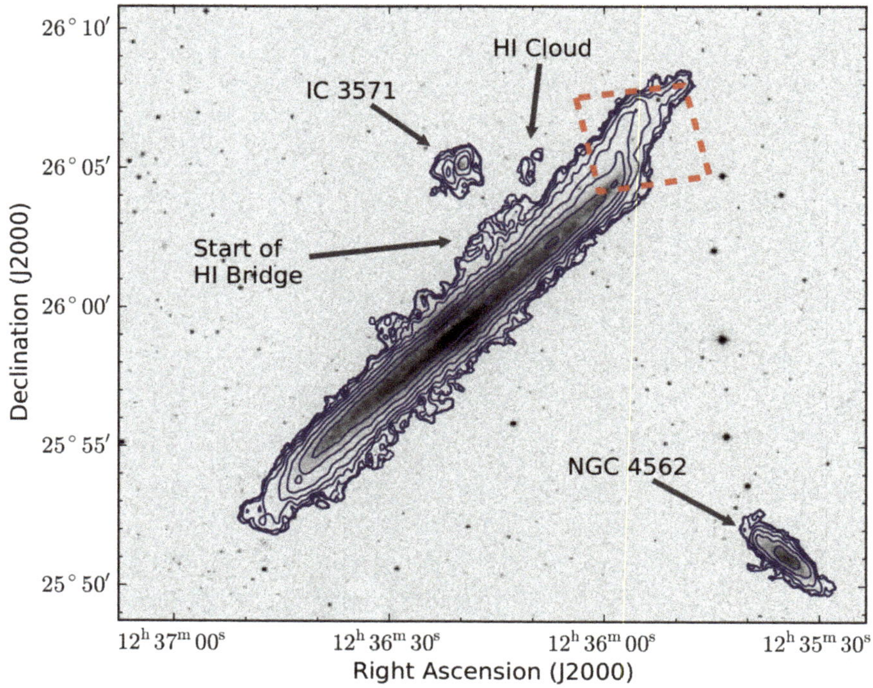

Figure 7.27. The H I observations of NGC 4565 (Zschaechner et al. 2012) with the field of view of Hubble overlaid. Several H I morphological elements are visible: two companion galaxies, IC 3571 and NGC 4562, an H I cloud above the disk of NGC 4565, and the beginning of the bridge of H I gas from NGC 4565 to IC 3571. The Hubble image is centered on the H I warp. The stars found in the Hubble image are further explored in Figure 7.28. Figure reproduced from Radburn-Smith et al. (2014).

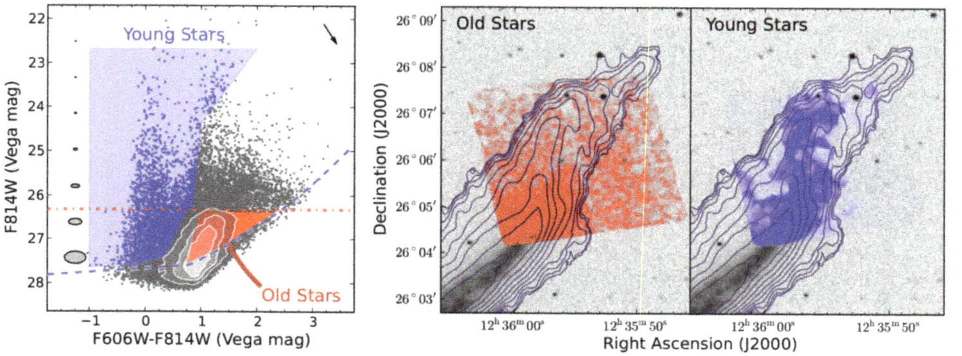

Figure 7.28. The color–magnitude diagram of all the stars found in the Hubble image of the warp region in NGC 4565 (red dashed box in Figure 7.27). The image has both young (main sequence) stars in blue and older (red giant branch) stars in red. The distribution of old stars suggests the old stellar disk truncates where the H I warp begins. The blue, younger stars follow the H I gas and were likely formed in the H I warp. An asymmetric warp with associated star formation is a possible indication of an accretion origin for this warp. Figure reproduced from Radburn-Smith et al. (2014).

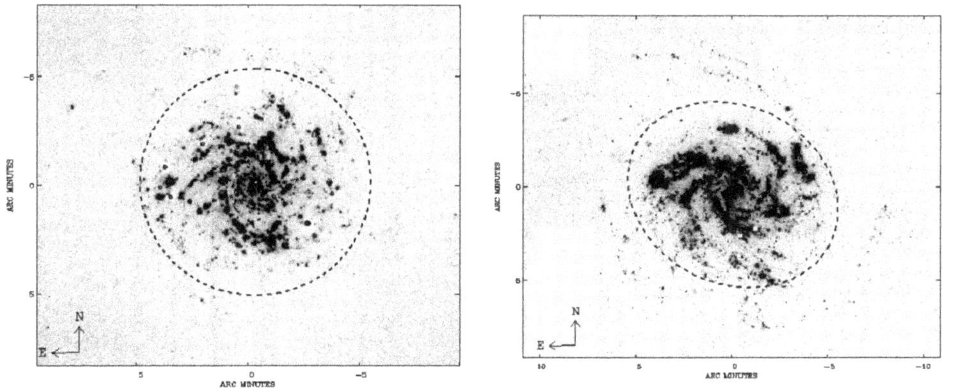

Figure 7.29. The first examples of star formation signatures found well outside what typically was thought as the "stellar disk" (the dashed line is the R_{25}). Left is NGC 628 and right NGC 6946. These are images taken in a narrow filter sensitive to $H\alpha$ emission line (ionized hydrogen), typically found near an ionizing stellar source such as UV emitting O- and B-stars. Figures reproduced from Ferguson (1998). © 1998. The American Astronomical Society. All rights reserved.

very edge of the stellar and H I disk. The older population (red, RGB stars) breaks at the end of the stellar disks and does not follow the warp. The younger He-burning main sequence population (blue) follows the warp however, suggesting star formation in the gas warp and possibly an accretion origin for this warp (Radburn-Smith et al. 2014; Gilhuly et al. 2020). Not only is the gas disk warped but, once imaged deeply enough, many stellar disks show signs of warping in the outermost radii (Saha et al. 2009).

7.7 XUV Disks

Interest in the outskirts of spiral galaxy disks has increased over recent years as these regions are the site of the most recent acquisition of gas for these systems (cf Sancisi et al. 2008), as well as low-level star formation (e.g., Dong et al. 2008; Bigiel et al. 2010; Alberts et al. 2011), making these faint outskirts the interface between the island universes—the galaxies themselves—and the cosmic web of primordial gas.

The low-level star formation was first discovered in Hα emission (Figure 7.29) by Ferguson et al. (1998) and Lelièvre & Roy (2000). After the launch of the Galaxy Evolution EXplorer (GALEX; Martin et al. 2005), initial anecdotal evidence pointed to ultraviolet disks of spiral galaxies extending much beyond their optical radius (Figure 7.30; Thilker et al. 2005b, 2005a; Gil de Paz et al. 2005, 2007; Zaritsky & Christlein 2007). The ultraviolet and extended Hα emission seemed closely related and extend for much of the atomic hydrogen disk (Figure 7.31).

Subsequent structural searches for these extended ultraviolet (XUV) disks by Thilker et al. (2007) and Lemonias et al. (2011) find that some 20%–30% of spirals posses an XUV disk and 40% of S0s (Moffett et al. 2012), making this type of disk common but not typical for spiral and S0 galaxies. These XUV disk complexes are

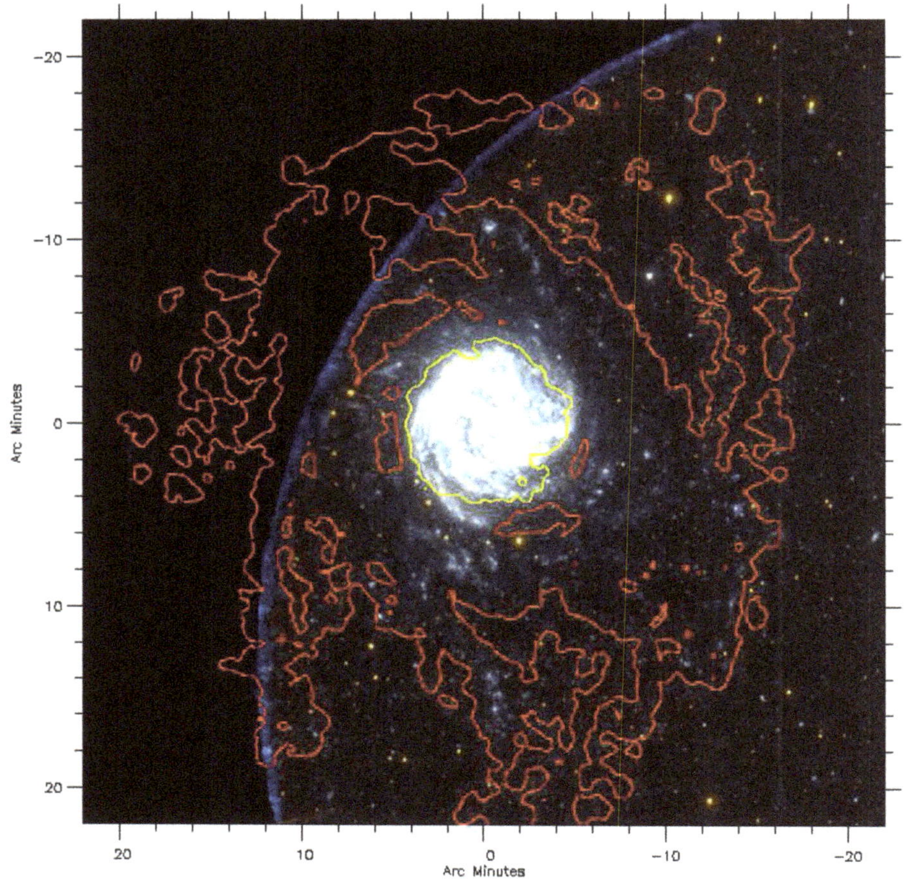

Figure 7.30. The first extended ultraviolet disk found in the face-on, nearby galaxy M83. Red contour is the atomic hydrogen map (HI) and the colors are ultraviolet light. Star formation is happening throughout the extended HI disk, not just in the central galaxy. Figure reproduced from Thilker et al. (2005b).

generally ~100 Myr old, explaining why most lack Hα (Alberts et al. 2011), as opposed to a top-light IMF[6] (as proposed by Meurer et al. 2009), and sub-solar but not excessively low metallicities (0.1–1 $Z_{\odot}$, based on emission lines; Gil de Paz et al. 2007; Bresolin et al. 2009; Werk et al. 2010). Additionally, it has been known for some time now that atomic hydrogen (H I) as observed by the 21 cm fine structure line also extends well beyond the optical disk of spiral galaxies (Begeman 1989; Meurer et al. 1996, 1998; Swaters et al. 2002; Noordermeer et al. 2005; Walter et al. 2008; Boomsma et al. 2008; Elson et al. 2011; Heald et al. 2011b, 2011a; Zschaechner et al. 2011). In those few cases where both high-quality H I and deep

[6] Initial mass function: the mass distribution typical for a new generation of stars: a few giants (O- and B-stars) and logarithmic steps with lower mass ending with hundreds of M-dwarf stars for every O-star born. Top-light means an even higher ratio in favor of lower-mass stars.

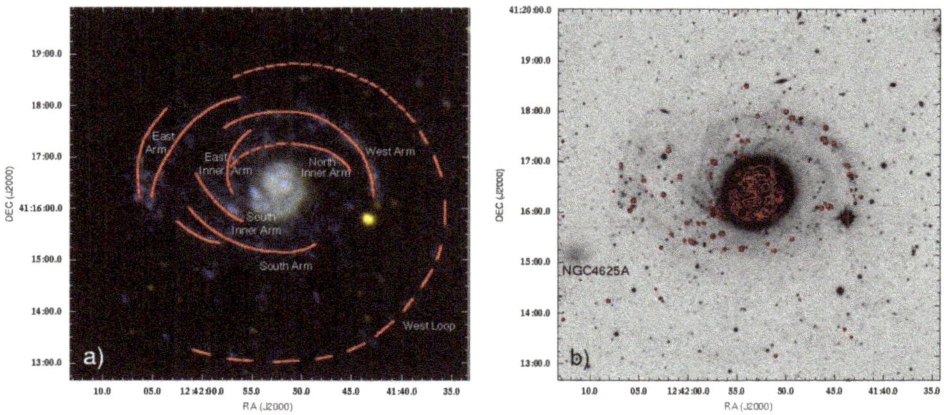

Figure 7.31. The XUV disk of NGC 4625 in the ultraviolet with the GALEX satellite (left panel) and the optical image with Hα marked in red (right panel). Not only do both Hα and UV light extend beyond the radius originally thought of as the edge of the stellar disk, both trace each other well. Figure reproduced from Gil de Paz et al. (2005). © 2005. The American Astronomical Society. All rights reserved.

GALEX data are available, a close relation in their respective morphology was remarked upon (Bigiel et al. 2010). While we have to wait for the all-sky surveys in H I to catch up to the coverage of the GALEX surveys (e.g., the wide survey with WSRT/APERTIF or the WALLABY survey with ASKAP), we can compare the morphology for those galaxies for which uniform H I information is available.

The canonical view of the origin of the XUV disks, is that the Kennicutt–Schmidt law (Kennicutt 1998) needs to be extended to low global surface densities of gas and the formation of individual O-stars in the very outskirts of disks (Cuillandre et al. 2001; Bigiel et al. 2010). The recent accretion of cold gas flows (Kereš et al. 2005) into the H I disk is the origin for the young stars (the fueling rate implied by XUV disks is explored; Lemonias et al. 2011) and H I warps (Roškar et al. 2010). The fraction of spirals that have an XUV disk (∼20%–30%) supports this scenario as the remaining spirals may simply have no current cool gas inflow. In this case, one would expect UV and H I morphology to follow each other reasonably closely for the XUV disks but not for many of the others, as the star formation in the inner disk is more closely related to the molecular phase (Bigiel et al. 2008).

However, the existence of XUV disks pose an intriguing alternate possibility for the origin of the atomic hydrogen disk. Instead of primordial gas accreting onto the disk, the H I disk could also be the byproduct of photodissociation of molecular hydrogen on the "skins" of molecular clouds by the ultraviolet flux of the young stars in the XUV disk (Allen et al. 1997, 2004; Allen 2002). This explanation has been explored in the stellar disks of several nearby galaxies (Heiner et al. 2008a, 2008b, 2009, 2010). One can calculate the timescales (molecular gas dissociation and re-formation) involved but these are inconclusive regarding the origin of the XUV disk (Gil de Paz et al. 2007). In this scenario, one would expect the UV and H I morphologies to follow each other closely in all cases; in the outer disk, the UV flux

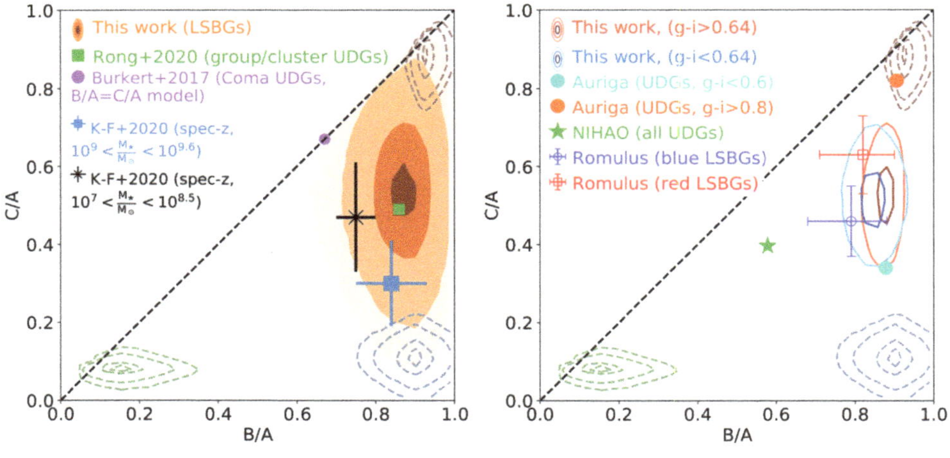

Figure 7.32. A compilation of UDG axes ratios (Kado-Fong et al. 2021). Left are the observed and then modeled axes ratios (Burkert 2017; Rong et al. 2020; Kado-Fong et al. 2020). Right are the results from cosmological simulations: Auriga (Liao et al. 2019) and NIHAO (Jiang et al. 2019). Figure reproduced from Kado-Fong et al. (2021). © 2021. The American Astronomical Society. All rights reserved.

from a few young stars would reach out to large areas of the low-column density gas to dissociate enough hydrogen to form the outer H I disk. Thus, the low-flux and low H I column density morphology—those defining the limits and extent of the XUV and H I disks—should show a close relation in parameters such as concentration, Gini, M_{20} and the effective radius (see Chapter 6). A third explanation is in terms of recent tidal interaction. A major merger often pulls gas out of the planes of galaxies and triggers star-forming events. Some anecdotal evidence (e.g., UGC 04862 is a late-stage major merger; Torres-Flores et al. 2012) does point to this possible origin. In this scenario, one would expect that most H I disks hosting an XUV disk would be seriously tidally disrupted.

7.8 LSB Dwarfs

Low surface brightness galaxies are often dwarf galaxies (in terms of stellar mass) residing in a dark matter halo, sometimes much more massive than one would expect for so few stars (i.e., a "failed galaxy" one that made only few stars for the dark matter residing there.)

7.8.1 Ultra-diffuse Galaxies

The recent discovery of a substantial number of very large diffuse galaxies in local clusters, has drawn significant attention to the possibility of LSB companions solving the "satellite problem" in recent years. van Dokkum et al. (2015) and Merritt et al. (2014) identified a significant population of large ($r_{\rm eff}$ > 1.5 kpc), low central surface brightness (μ_g = 24–26 mag arcsec^{-2}) galaxies in the Coma cluster, and dubbed these ultra diffuse galaxies (UDGs).

In the years since, several studies reported populations of UDGs in different galaxy cluster samples. Some of these focus on the nearby and well-studied Fornax (Muñoz-Mateos et al. 2015), Coma (Koda et al. 2015; Kadowaki et al. 2017), and Virgo (Mihos et al. 2015; Davies et al. 2016) clusters. Others studied more distant clusters (van der Burg et al. 2016; Roman & Trujillo 2017), including a Frontier-fields cluster at $z = 0.3$ (Janssens et al. 2017).

It is clear that UDGs are a common phenomenon in galaxy clusters, and their ubiquity has to be understood in the context of existing galaxy formation models. At present they either represent a new end-point of galaxy evolution, a failed satellite and tantalizingly a population of galaxies dominated by dark matter (van Dokkum et al. 2016), or severely lacking dark matter (van Dokkum et al. 2018), or both.

From integrated light, one can reconstruct the axes ratio (Figure 2.7) for these UDGs and answer the basic question: are these spheroids or disks? Figure 7.32 shows the results from observations and simulated galaxies. UDGs span a range of shapes from almost disks to spheroids.

7.8.2 Low Surface Brightness from Star Counts

The Antlia Dwarf galaxy is a satellite of our Milky Way. Many of the smaller satellites are now discovered or characterized as collections of individual stars. It becomes difficult to translate a loose grouping of stars, selected by their commonality in color–magnitude space.

Figure 7.33 shows the Antlia Dwarf (Pimbblet & Couch 2012) where there is a density map of the RGB stars (see Figure 7.9) from HST observations. To convert this count into a galaxy morphology, one can assign area to each star in the selection, either by binning the counts (which is very susceptible to incompleteness i.e., missing a few stars because of an artifact or a bright foreground star) or one can do so more dynamically using tricks like Voronoi tessellations. Beware that how and where one assigns area to the star count affects the measures shape pretty definitively.

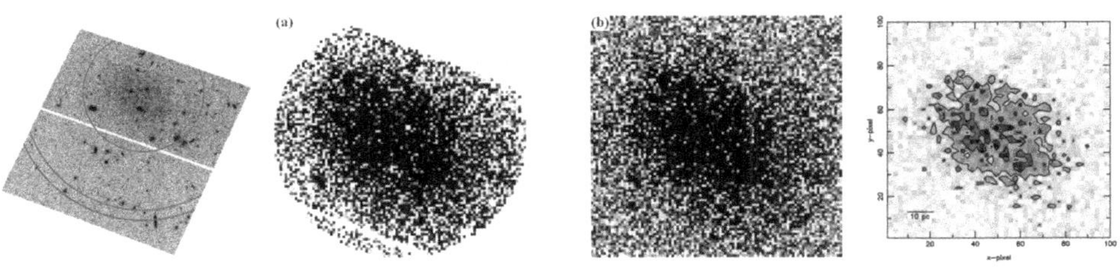

Figure 7.33. The Antlia dwarf galaxy, observed with Hubble Space Telescope Advanced Camera for Surveys (ACS) Left: the raw image, replete with background galaxies, chip gap and blue circles indicating the radii of the galaxy and the cutout for the next panel. Middle panels: panel a, the RGB star counts converted into a grayscale image, panel b, the same but with substitutions for where information is missing. Right: the star count converted to a surface density by binning the data. Star counts can be converted into an image of "integrated light" but only after converting the number densities to a surface brightness and correcting for any missing counts. Figures reproduced from Pimbblet & Couch (2012). © 2012 The Authors Monthly Notices of the Royal Astronomical Society © 2012 RAS.

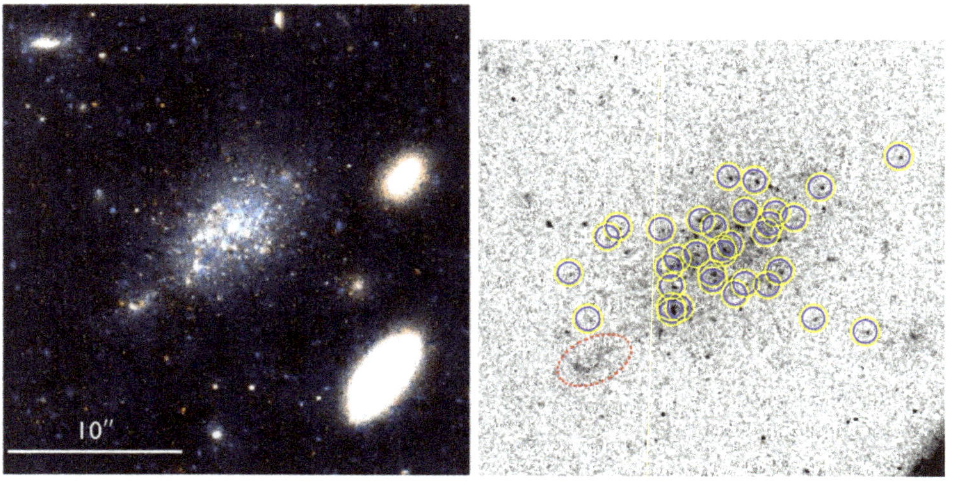

Figure 7.34. Left panel: color image of GHOSTS-I, in one of the GHOSTS fields located at a projected distance of 32 kpc from M81. Right panel: F606W-band image of GHOSTS-I with circles indicate the individual stars in the region of the dwarf detected using the GHOSTS pipeline, and the red dashed ellipse indicates a possible star-forming region. Figure reproduced from Monachesi et al. (2014). © 2014. The American Astronomical Society. All rights reserved.

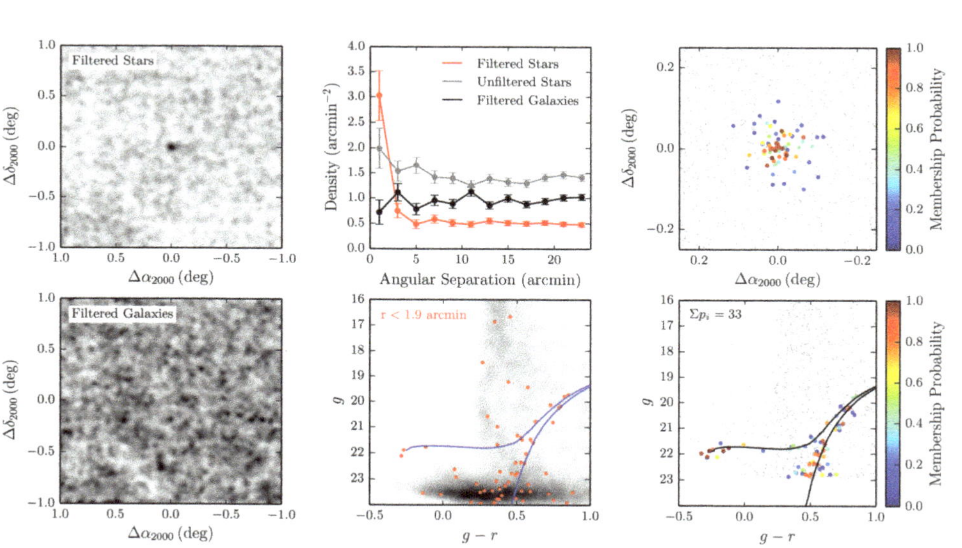

Figure 7.35. DES J0531–2801 (Columba I) Top left: spatial distribution of stellar objects with $g < 24AB$ mag. The field of view is $2° \times 2°$ centered on the candidate. Top center: radial distribution of objects with $g - r < 1$ mag and $g < 24AB$ mag: selected stars (red), other stars (gray), and selected galaxies (black). Top right: spatial distribution of stars with high membership probabilities within a $0.5° \times 0.5°$ field. Bottom left: same as top left panel, but for probable galaxies that pass the selection. Bottom center: the color–magnitude distribution of stars within of the centroid are indicated with individual points, field objects in gray. The blue curve shows the best-fit isochrone. Bottom right: color–magnitude distribution of high membership probability stars. The identification of such very low mass satellite galaxies of our Milky Way has now almost been completed with a large number identified in this way. Figure reproduced from Drlica-Wagner et al. (2015). © 2015. The American Astronomical Society. All rights reserved.

Although it is rare, given how small an area of sky the Hubble has looked so far, it has discovered some LSB dwarfs by accident: GHOSTS-I and -II resolved out from a faint smudge on Sloan images to a small number of stars (Figure 7.34). Nearer to the Milky Way, the DES collaboration found dwarfs as over-densities in the star counts of the Galactic Halo. With future missions such as the Roman and Euclid Space Telescopes, there will be deep, wide and Hubble-sharp images to look for over-densities of resolved stars (Figure 7.35).

7.9 Assignments

There one assignment to go with low surface brightness galaxies: "Dark Matter in Galaxies" (Jupyter notebook 11) estimating the dark matter content of a ultra-diffuse galaxy from van Dokkum et al. (2019a).

7.10 Further Reading

Disk Warps and Truncations the review on disk galaxies by van der Kruit (van der Kruit & Freeman 2011) is excellent.

LSB Integrated Light there are two excellent review papers out on the subject of integrated light: Knapen & Trujillo (2017) and Martinez-Delgado (2020). The first one for most of the phenomena discussed here and how to observe them in integrated light, the second on how to do so with small "amateur" telescopes.[7]

LSB from star counts there are two approaches here, the unambiguous Hubble star counts and the much more challenging ground-based star counts. The latter can be done successfully on nearby galaxies, famously M31. Further reading on that is the PanDAS survey.

References

Abraham, R., Merritt, A., Zhang, J., et al. 2016, in IAU Proc. 11, 137

Abraham, R. G., & van Dokkum, P. G. 2014, PASP, 126, 55

Alberts, S., Calzetti, D., Dong, H., et al. 2011, ApJ, 731, 28

Allen, R. J. 2002, in ASP Conf. Ser. 276, Seeing Through the Dust: The Detection of HI and the Exploration of the ISM in Galaxies (San Francisco, CA: ASP), 288

Allen, R. J., Heaton, H. I., & Kaufman, M. J. 2004, ApJ, 608, 314

Allen, R. J., Knapen, J. H., Bohlin, R., & Stecher, T. P. 1997, ApJ, 487, 171

Aumer, M., Binney, J., & Schönrich, R. 2016, MNRAS, 462, 1697

Bailin, J., & Harris, W. E. 2009, ApJ, 695, 1082

Bakos, J., & Trujillo, I. 2012, arXiv:1204.3082

Begeman, K. G. 1989, A&A, 223, 47

Bell, E. F., Monachesi, A., Harmsen, B., et al. 2017, ApJL, 837, L8

Belokurov, V., Zucker, D. B., Evans, N. W., et al. 2006, ApJL, 642, L137

[7] Not a fan of the term "amateur astronomer" since in English it seems to convey lower quality work. Observational astronomers know that Astronomer Aficionados can produce outstanding data. These deep images are an example of such outstanding data. Other examples are stellar photometry and exoplanet hunting.

Beraldo e Silva, L., Debattista, V. P., Khachaturyants, T., & Nidever, D. 2020, MNRAS, 492, 4716
Bigiel, F., Leroy, A., Seibert, M., et al. 2010, ApJL, 720, L31
Bigiel, F., Leroy, A., Walter, F., et al. 2008, AJ, 136, 2846
Boomsma, R., Oosterloo, T. A., Fraternali, F., van der Hulst, J. M., & Sancisi, R. 2008, A&A, 490, 555
Bovy, J., Rix, H.-W., Green, G. M., Schlafly, E. F., Finkbeiner, D. P., et al. 2016, ApJ, 818, 130
Bovy, J., Rix, H.-W., Hogg, D. W., et al. 2012, ApJ, 755, 115
Bresolin, F., Ryan-Weber, E., Kennicutt, R. C., & Goddard, Q. 2009, ApJ, 695, 580
Briggs, F. H. 1990, ApJ, 352, 15
Bullock, J. S., & Johnston, K. V. 2005, ApJ, 635, 931
Bullock, J. S., Kravtsov, A. V., & Weinberg, D. H. 2001, ApJ, 548, 33
Burkert, A. 2017, ApJ, 838, 93
Comerón, S., Elmegreen, B. G., Knapen, J. H., et al. 2011a, ApJ, 741, 28
Comerón, S., Elmegreen, B. G., Salo, H., et al. 2014, A&A, 571, A58
Comerón, S., Knapen, J. H., Sheth, K., et al. 2011b, ApJ, 729, 18
Comerón, S., Salo, H., Janz, J., Lauikainen, E., & Yaochim, P. 2015, A&A, 584, A34
Comerón, S., Salo, H., & Knapen, J. H. 2018, A&A, 610, A5
Comerón, S., Salo, H., Knapen, J. H., & Peletier, R. F. 2019, A&A, 623, A89
Comerón, S., Salo, H., Peletier, R. F., & Mentz, J. 2016, A&A, 593, L6
Cooper, A. P., D'Souza, R., Kauffmann, G., et al. 2013, MNRAS, 434, 3348
Cooper, M. C., Coil, A. L., Gerke, B. F., et al. 2010, MNRAS, 409, 337
Cuillandre, J., Lequeux, J., Allen, R. J., Mellier, Y., & Bertin, E. 2001, ApJ, 554, 190
Danieli, S., van Dokkum, P., Merritt, A., et al. 2017, ApJ, 837, 136
Danieli, S., van Dokkum, P., & Conroy, C. 2018, ApJ, 856, 69
Davies, J. I., Davies, L. J. M., & Keenan, O. C. 2016, MNRAS, 456, 1607
de Jong, R. S. 2008, MNRAS, 388, 1521
Deason, A. J., Mao, Y.-Y., & Wechsler, R. H. 2016, ApJ, 821, 5
Dong, H., Calzetti, D., Regan, M., et al. 2008, AJ, 136, 479
Dotter, A., Chaboyer, B., Jevremovic, D., et al. 2008, ApJS, 178, 89
Driver, S. P., Norberg, P., Baldry, I. K., et al. 2009, A&G, 50, 5.12
Drlica-Wagner, A., Bechtol, K., Rykoff, E. S., et al. 2015, ApJ, 813, 109
D'Souza, R., & Bell, E. F. 2018, MNRAS, 474, 5300
Elmegreen, B. G., Elmegreen, D. M., Tompkins, B., & Jenks, L. G. 2017, ApJ, 847, 14
Elson, E. C., de Blok, W. J. G., & Kraan-Korteweg, R. C. 2011, MNRAS, 415, 323
Ferguson, A. M. N., Irwin, M. J., Ibata, R. A., et al. 2002, AJ, 124, 1452
Ferguson, A. M. N., Wyse, R. F. G., Gallagher, J. S., & Hunter, D. A. 1998, ApJL, 506, L19
Ferguson, H. C. 1998, in Space Telescope Science Institute Symp. 11, The Hubble Deep Field (New York: Cambridge Univ. Press), 181
Font, A. S., Johnston, K. V., Bullock, J. S., & Robertson, B. E. 2006, ApJ, 646, 886
Font, A. S., Johnston, K. V., Ferguson, A. M. N., et al. 2008, ApJ, 673, 215
García-Ruiz, I., Sancisi, R., & Kuijken, K. 2002, A&A, 394, 769
Gil de Paz, A., Madore, B. F., Boissier, S., et al. 2005, ApJL, 627, L29
Gil de Paz, A., Madore, B. F., Boissier, S., et al. 2007, ApJ, 661, 115
Gilhuly, C., Hendel, D., Merritt, A., et al. 2020, ApJ, 897, 108
Guérou, A., Emsellem, E., Krajnovic, D., et al. 2016, A&A, 591, A143

Harmsen, B., Monachesi, A., Bell, E. F., et al. 2017, MNRAS, 466, 1491
Haywood, M., Di Matteo, P., Lehnert, M. D., Katz, D., & Gomez, A. 2013, A&A, 560, A109
Heald, G., et al. 2011a, in IAU Symp. 277, 59
Heald, G., Jozsa, G., Serra, P., et al. 2011b, A&A, 526, A118
Heiner, J. S., Allen, R. J., Emonts, B. H. C., & van der Kruit, P. C. 2008a, ApJ, 673, 798
Heiner, J. S., Allen, R. J., van der Kruit, P. C., et al. 2009, in Proc. 4th Spitzer Science Center Conf., The Evolving ISM in the Milky Way and Nearby Galaxies, ed. K. Sheth, A. Noriega-Crespo, J. Ingalls, & R. Paladini, E28
Heiner, J. S., Allen, R. J., & van der Kruit, P. C. 2010, ApJ, 719, 1244
Heiner, J. S., Allen, R. J., Wong, O. I., & van der Kruit, P. C. 2008b, A&A, 489, 533
Helmi, A. 2020, ARA&A, 58, 205
Helmi, A., Babusiaux, C., Koppelman, H. H., et al. 2018, Natur, 563, 85
Herpich, J., Stinson, G. S., Dutton, A. A., et al. 2015, MNRAS, 448, L99
Huxor, A. P., Mackey, A. D., Ferguson, A. M. N., et al. 2014, MNRAS, 442, 2165
Ibata, R., Irwin, M., Lewis, G. F., & Stolte, A. 2001, ApJL, 547, L133
Jang, I. S., de Jong, R. S., Holwerda, B. W., et al. 2020, A&A, 637, A8
Janssens, S., Abraham, R., Brodie, J., et al. 2017, ApJ, 839, L17
Jiang, F., Dekel, A., Freundlich, J., et al. 2019, MNRAS, 487, 5272
Johnston, S., Taylor, R., Balies, M., et al. 2008, ExA, 22, 151
Kado-Fong, E., Greene, J. E., Huang, S., et al. 2020, ApJ, 900, 163
Kado-Fong, E., Petrescu, M., Mohammad, M., et al. 2021, ApJ, 920, 72
Kadowaki, J., Zaritsky, D., & Donnerstein, R. L. 2017, ApJL, 838, L21
Kasparova, A. V., Katkov, I. Y., & Chilingarian, I. V. 2020, MNRAS, 493, 5464
Kasparova, A. V., Katkov, I. Yu., Chilingarian, I. V., et al. 2016, MNRAS, 460, L89
Katkov, I. Y., Kniazev, A. Y., Kasparova, A. V., & Sil'chenko, O. K. 2019, MNRAS, 483, 2413
Kawata, D. 2018, in IAU Symp. 334, Rediscovering our Galaxy (Cambridge: Cambridge Univ. Press), 132
Kazantzidis, S., Zentner, A. R., Kravtsov, A. V., Bullock, J. S., & Debattista, V. P. 2009, ApJ, 700, 1896
Kennicutt, R. C. Jr 1989, ApJ, 344, 685
Kennicutt, R. C. Jr 1998, ApJ, 498, 541
Kereš, D., Katz, N., Weinberg, D. H., & Dave, R. 2005, MNRAS, 363, 2
Klypin, A., Kravtsov, A. V., Valenzuela, O., & Prada, F. 1999, ApJ, 522, 82
Knapen, J. H., & Trujillo, I. 2017, in Astrophysics and Space Science Library, Vol. 434, Outskirts of Galaxies (Berlin: Springer), 255
Koda, J., Yagi, M., Yamanoi, H., & Komiyama, Y. 2015, ApJ, 807, L2
Kruijssen, J. M. D., Pfeffer, J. L., Reina-Campos, M., Crain, R. A., & Bastian, N. 2019, MNRAS, 486, 3180
Laine, J., Laurikainen, E., Salo, H., et al. 2014, MNRAS, 441, 1992
Lelièvre, M., & Roy, J-R 2000, AJ, 120, 1306
Lemonias, J. J., Schiminovich, D., Thilker, D., et al. 2011, ApJ, 733, 74
Liao, S., Gao, L., Frenk, C. S., et al. 2019, MNRAS, 490, 5182
Lima Neto, G. B., Durret, F., Lagana, T. F., et al. 2020, A&A, 641, A95
Macciò, A. V., Moore, B., Stadel, J., et al. 2006, ApJL, 636, L25
Mackey, A. D., Huxor, A. P., Ferguson, A. M. N., et al. 2010, ApJL, 717, L11
Majewski, S. R., Skrutskie, M. F., Weinberg, M. D., & Ostheimer, J. C. 2003, ApJ, 599, 1082

Malin, D. F. 1977, AAS Photo Bulletin, #16, 10
Malin, D. F., & Carter, D. 1983, ApJ, 274, 534
Marsakov, V. A., Koval', V. V., & Gozha, M. L. 2020, AstBu, 75, 21
Martig, M., Minchev, I., & Flynn, C. 2014, MNRAS, 442, 2474
Martig, M., Minchev, I., Ness, M., Fouesneau, M., & Rix, H.-W. 2016, ApJ, 831, 139
Martin, D. C., Fanson, J., Schiminovich, D., et al. 2005, ApJL, 619, L1
Martin, N. F., McConnachie, A. W., Irwin, M., et al. 2009, ApJ, 705, 758
Martinez-Delgado, D. 2019, in Proc. XIII Scientific Meeting of Spanish Astronomical Society, Highlights on Spanish Astrophysics X, 146
Martinez-Delgado, D. 2020, arXiv:2001.05746
Martinez-Delgado, D., Cooper, A. P., Roman, J., et al. 2021a, arXiv:2104.06071
Martinez-Delgado, D., D'Onghia, E., Chonis, T. S., et al. 2015, AJ, 150, 116
Martínez-Delgado, D., Gabany, R. J., Crawford, K., et al. 2010, AJ, 140, 962
Martínez-Delgado, D., Lasker, R., Sharina, M., et al. 2016, AJ, 151, 96
Martínez-Delgado, D., Pohlen, M., Gabany, R. J., et al. 2009, ApJ, 692, 955
Martinez-Delgado, D., Roman, J., Erkal, D., et al. 2021b, MNRAS, 506, 5030
Martínez-Delgado, D., Romanowsky, A. J., Gabany, R. J., et al. 2012, ApJL, 748, L24
Martínez-Lombilla, C., & Knapen, J. H. 2019, A&A, 629, A12
Martínez-Lombilla, C., Trujillo, I., & Knapen, J. H. 2019, MNRAS, 483, 664
McConnachie, A. W., Ibata, R., Martin, N., et al. 2018, ApJ, 868, 55
McConnachie, A. W., Irwin, M. J., Ibata, R. A., et al. 2009, Natur, 461, 66
Merritt, A., van Dokkum, P., Abraham, R., et al. 2014, ApJL, 787, L37
Merritt, A., van Dokkum, P., Abraham, R., & Zhang, J. 2016a, ApJ, 830, 62
Merritt, A., van Dokkum, P., Danieli, S., et al. 2016b, ApJ, 833, 168
Meurer, G. R., Carignan, C., Beaulieu, S. F., & Freeman, K. C. 1996, AJ, 111, 1551
Meurer, G. R., Staveley-Smith, L., & Killeen, N. E. B. 1998, MNRAS, 300, 705
Meurer, G. R., Wong, O. I., Kim, J. H., et al. 2009, ApJ, 695, 765
Mihos, C., Durrell, P. R., Ferrarese, L., et al. 2015, ApJ, 809, L21
Mihos, J. C., Harding, P., Spengler, C. E., Rudick, C. S., & Feldmeier, J. J. 2013, ApJ, 762, 82
Minchev, I., Famaey, B., Quillen, A. C., et al. 2012, A&A, 548, A126
Minchev, I., Martig, M., Streich, D., et al. 2015, ApJ, 804, L9
Moffett, A. J., Kannappan, S. J., Baker, A. J., & Laine, S. 2012, ApJ, 745, 34
Monachesi, A., et al. 2014, ApJ, 780, 179
Monachesi, A., Bell, E. F., Radburn-Smith, D. J., et al. 2013, ApJ, 766, 106
Monachesi, A., Bell, E. F., Radburn-Smith, D. J., et al. 2016, MNRAS, 457, 1419
Mosenkov, A., Rich, R. M., Koch, A., et al. 2020a, MNRAS, 494, 1751
Mosenkov, A. V., Smirnov, A. A., Sil'chenko, O. K., et al. 2020b, MNRAS, 497, 2039
Mouhcine, M., Ibata, R., & Rejkuba, M. 2010, ApJL, 714, L12
Muñoz-Mateos, J. C., Sheth, K., Regan, M., et al. 2015, ApJS, 219, 3
Noordermeer, E., van der Hulst, J. M., Sancisi, R., Swaters, R. A., & van Albada, T. S. 2005, A&A, 442, 137
Peters, S. P. C., van der Kruit, P. C., Knapen, J. H., et al. 2017, MNRAS, 470, 427
Pimbblet, K. A., & Couch, W. J. 2012, MNRAS, 419, 1153
Pinna, F., Falcon-Barroso, J., Martig, M., et al. 2019a, A&A, 625, A95
Pinna, F., Falcon-Barroso, J., Martig, M., et al. 2019b, A&A, 623, A19

Purcell, C. W., Bullock, J. S., & Kazantzidis, S. 2010, MNRAS, 404, 1711
Purcell, C. W., Bullock, J. S., & Zentner, A. R. 2007, ApJ, 666, 20
Qu, Y., Di Matteo, P., Lehnert, M. D., & van Driel, W. 2011, A&A, 530, A10
Radburn-Smith, D. J., de Jong, R. S., Seth, A. C., et al. 2011, ApJS, 195, 18
Radburn-Smith, D. J., de Jong, R. S., Streich, D., et al. 2014, ApJ, 780, 105
Rich, R. M., Brosch, N., Bullock, J., et al. 2016, in IAU Proc. 11, 186
Rich, R. M., Brosch, N., Bullock, J., et al. 2017, in IAU Symp. 321, Formation and Evolution of Galaxy Outskirts (Cambridge: Cambridge Univ. Press), 186
Roman, M., & Trujillo, I. 2017, MNRAS, 468, 4039
Rong, Y., Mancera Piña, P. E., Tempel, E., Puzia, T. H., & De Rijcke, S. 2020, MNRAS, 498, L72
Roškar, R., Debattista, V. P., Brooks, A. M., et al. 2010, MNRAS, 408, 783
Roškar, R., Debattista, V. P., Quinn, T. R., & Wadsley, J. 2012, MNRAS, 426, 2089
Saha, K., de Jong, R., & Holwerda, B. 2009, MNRAS, 396, 409
Sancisi, R., Fraternali, F., Oosterloo, T., & van der Hulst, T. 2008, A&ARv, 15, 189
Schweizer, F., & Ford, W. K. J. 1985, in Lecture Notes in Physics, Vol. 232, New Aspects of Galaxy Photometry, ed. J. L. Nieto (Berlin: Springer), 145
Schweizer, F., Whitmore, B. C., & Rubin, V. C. 1983, AJ, 88, 909
Snaith, O. N., Haywood, M., Di Matteo, P., et al. 2014, ApJL, 781, L31
Spavone, M., Iodice, E., Arnaboldi, M., et al. 2010, ApJ, 714, 1081
Streich, D., de Jong, R. S., Baillin, J., et al. 2016, A&A, 585, A97
Swaters, R. A., van Albada, T. S., van der Hulst, J. M., & Sancisi, R. 2002, A&A, 390, 829
Thilker, D. A., Bianchi, L., Boissier, S., et al. 2005a, ApJL, 619, L79
Thilker, D. A., Bianchi, L., Meurer, G., et al. 2007, ApJS, 173, 538
Thilker, D. A., Hoopes, C. G., Bianchi, L., et al. 2005b, ApJL, 619, L67
Torres-Flores, S., de Oliveira, C. M., de Mello, D. F., Scarano, S., & Urrutia-Viscarra, F. 2012, MNRAS, 421, 3612
Trujillo, I. 2016, in IAU Proc. 11, 155
van der Burg, R. F. J., Muzzin, A., & Hoekstra, H. 2016, A&A, 590, A20
van der Kruit, P. C. 1987, A&A, 173, 59
van der Kruit, P. C. 2007, A&A, 466, 883
van der Kruit, P. C., & Freeman, K. C. 2011, ARA&A, 49, 301
van der Kruit, P. C., & Searle, L. 1981, A&A, 95, 105
van der Kruit, P. C., & Searle, L. 1982, A&A, 110, 79
van Dokkum, P., Abraham, R., Brodie, J., et al. 2016, ApJL, 828, L6
van Dokkum, P., Danieli, S., Cohen, Y., et al. 2018, Natur, 555, 629
van Dokkum, P., Danieli, S., Abraham, R., Conroy, C., & Romanowsky, A. J. 2019a, ApJ, 874, L5
van Dokkum, P., Gilhuly, C., Bonaca, A., et al. 2019b, ApJL, 883, L32
van Dokkum, P. G., Abraham, R., & Merritt, A. 2014, ApJL, 782, L24
van Dokkum, P. G., Abraham, R., Merritt, A., et al. 2015, ApJL, 798, L45
Walter, F., Brinks, E., de Blok, W. J. G., et al. 2008, AJ, 136, 2563
Werk, J. K., Putman, M. E., Meurer, G. R., et al. 2010, ApJ, 715, 656
White, S. D. M., & Rees, M. J. 1978, MNRAS, 183, 341
Yoachim, P., & Dalcanton, J. J. 2005, ApJ, 624, 701

Yoachim, P., & Dalcanton, J. J. 2006, AJ, 131, 226
Yoachim, P., & Dalcanton, J. J. 2008a, ApJ, 683, 707
Yoachim, P., & Dalcanton, J. J. 2008b, ApJ, 682, 1004
Younger, J. D., Cox, T. J., Seth, A. C., & Hernquist, L. 2007, ApJ, 670, 269
Yu, S., Bullock, J. S., Courtney, K., et al. 2021, MNRAS, 505, 889
Zaritsky, D., & Christlein, D. 2007, AJ, 134, 135
Zhang, J., Abraham, R., van Dokkum, P., Merritt, A., & Janssens, S. 2018, ApJ, 855, 78
Zschaechner, L. K., Rand, R. J., Heald, G. H., Gentile, G., & Józsa, G. 2012, ApJ, 760, 37
Zschaechner, L. K., Rand, R. J., Heald, G. H., Gentile, G., & Kamphuis, P. 2011, ApJ, 740, 35

Chapter 8

Kinematics and Morphology

The orbits of stars are in a large part what shapes the morphology of the galaxy they are in. There are some excellent books on galaxy kinematics and the goal is not to redo those here but to give an overview of the kind of processes thought to be linked to each morphological feature.

8.1 Evolution: Random to Rotation to Random

Galaxy evolve kinematically as well as in stellar content (see Chapter 10) and the main evolution is that they start (likely) from random motions (as the original cloud of material falls into the dark matter potential), eventually settling into a rotating disk, which in turn is randomized by encounters with other galaxies (mergers) into new randomized motion.

There are two terms associated with galaxy orbits to learn here: "pressure supported" and "rotationally supported." The first is a mix of randomly aligned orbits and the second is the rotating disk. What kind of motion (kinetic energy) do the stars have so that they do not sink further into the potential well of the dark matter?

8.1.1 Disk Formation

In general terms, disks of galaxies are formed when the initial cloud they collapse from has a tiny bit of net rotation. This initial angular momentum vector defines a plane of rotation and the direction of motion for the cloud at it shrinks substantially. In this plane, the rotation speeds up as mass falls in (conservation of momentum). In the vertical dimension, the infall results in random collisions and eventually these cancel out and the material settles into a thin disk. We can see this in the formation of disk galaxies or planetary system. It explains why it all rotates in one direction in a thin plane. It also strongly implies the system formed from a primordial cloud.

doi:10.1088/2514-3433/ac2c7dch8

8.1.2 Rotation Back to Pressure Supported

Once the gas and stars from it are organized in a disk, this represents a system with a clear angular momentum direction. The distributions of these angular momentum orientations seem to be random throughout the universe (the universe has no preferred direction of rotation which would have been weird but *interesting*) but may be aligned weakly with larger structure e.g., the filaments and walls of galaxies.

It would take a lot of work to change or randomize the angular momentum of so many stars. This happens in two ways: one is disk heating. The random encounters between stars or with large molecular clouds or a tidal encounter (more later) add some randomization to their orbit. This "heating" of the disk means the disk becomes thicker and this is prime candidate for both pseudo-bulges (Chapter 4) and thick disks (Chapter 7).

More radically, two disk galaxies can smash together. We see in the local universe examples of two disk galaxies with a myriad of tidal features. Even the earliest simulations of such interactions showed that the end product meant so much angular momentum would be exchanged that all the stellar orbits ended up completely randomized: the result is an elliptical galaxy. Such an outcome is not guaranteed and even with major merger a disk galaxy can still be the outcome depending on the initial galaxy orbits.

8.2 Rotating Disks

Disk galaxies rotate around their center. The rotation is followed by both the stars and the gas in the disk. The rotation speed rises quickly measured from the center of the galaxy outward and settles on a fairly constant value for most of the remainder of the disk. Some disks still rise some with radius, some "turn over" and decline a little. This rotation can be explained as orbits around the total mass enclosed by the orbit. If one assumes all the mass in a galaxy is associated with the stars, like it is with our solar system (98% of the mass is in the Sun), the rotation velocity in the outskirts should decline (think Kepler's laws). However, since the rotation speed remains relatively flat, another component needs to be contributing mass in the outer parts of the disk of galaxies: this was the first evidence of dark matter (Bosma et al. 1977; Rubin 1980).

In this context, the morphology is important to identify the stellar components in order to constrain the dark matter contribution. The bulge/disk decompositions (e.g. Figure 8.1) and accurate characterization is to account for the mass which is *not* mysterious.

8.3 Vertical Stability

The surface gravity of all the gas and stars in the disk of galaxies influences how the dust clouds are arranged. Dalcanton et al. (2004) noted how massive disks appear to have a continuous dark dust lane in the plane of the disk and less-massive edge-on disk galaxies do not (Figure 8.2). Yet there is evidence for dust in both kinds of galaxies. There are dark clouds visible, the stars are reddened by dust etc.

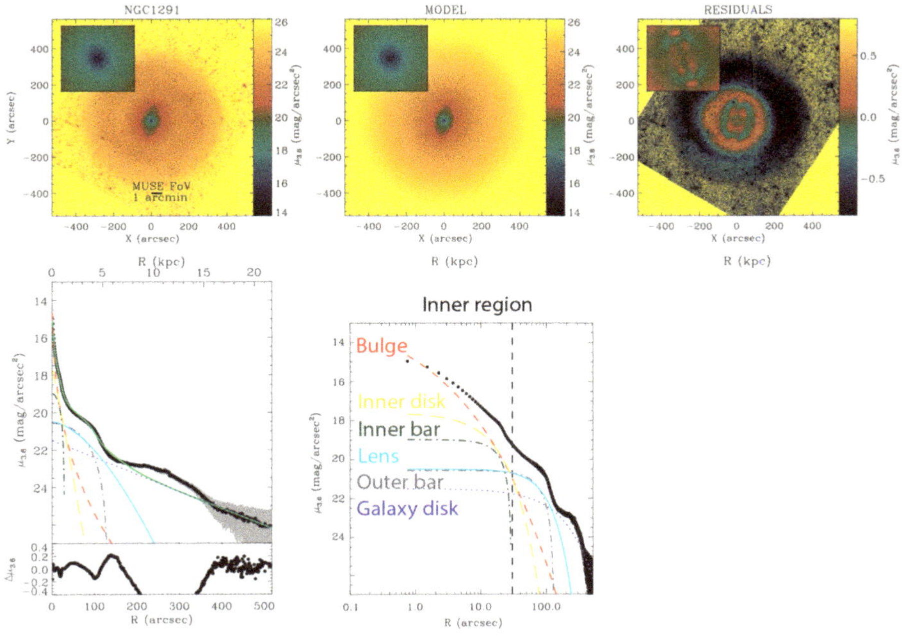

Figure 8.1. 2D multi-component photometric decomposition performed with GASP2D for the double-barred galaxy NGC 1291. The top panels show the S4G image, the best-fitting 2D model, and the model-image residuals (from left to right). The insets zoom inside the corresponding MUSE FoV. The bottom-left panel shows the radial profile of the image with black dots and the same measurement for the model overplotted as a green line. This galaxy has been fitted with 6 structural components: bulge (red dashed line), inner bar (dark-green dash–dotted line), inner disk (yellow dashed line), outer bar (gray dash–dotted line), lens (cyan dash–triple-dotted line), and galaxy disk (blue dotted line). The bottom right shows a zoom around the central regions, with the extent of the MUSE FoV indicated by a vertical dashed black line. Figure reproduced from de Lorenzo-Cáceres et al. (2019). © 2019 The Author(s), Published by Oxford University Press on behalf of the Royal Astronomical Society.

Computing the vertical stability ($1/Q$) of the radial mass profile of these disks (Figure 8.3) from the kinematics and surface density profile, it becomes clear that the inner parts of massive (v > 120 km/s rotation speed) disk galaxies is *unstable*, meaning that the timescale is short for any cloud that out of the disk (e.g., supernovae) to return to mid-plane. This imbalance between turbulence and gravity resulting in a dust disk in massive galaxies has been borne out by the Galaxy Zoo (Holwerda et al. 2019) and to higher redshifts (Figure 10.5; Holwerda et al. 2012). Not only is this effect noticeable on disposition of dust clouds in edge-on disks, the stellar disks themselves are relatively flatter (in the infrared): Figure 8.4 shows how the ratio of radial and vertical scales flattens out for more massive (rotation velocity above v = 120 km s^{-1}).

8.4 Bars

The TIMER project explores the centers of barred galaxies in much greater detail thanks to Integral Field Unit (IFU) observations. In essence, each pixel in an IFU

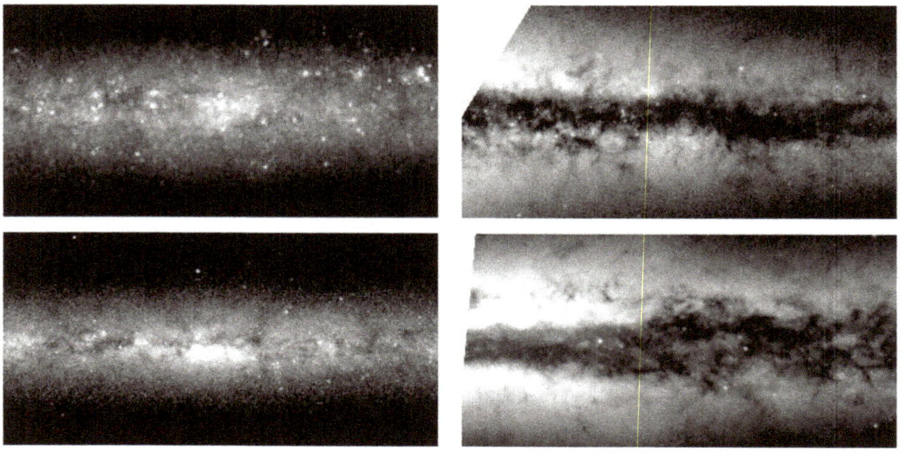

Figure 8.2. Examples of edge-on galaxies with low-mass galaxies left and high-mass galaxies right. Figure reproduced from Dalcanton et al. (2004).

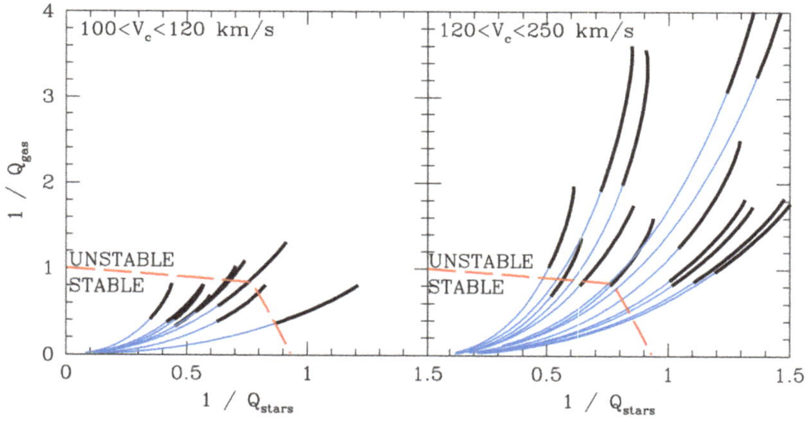

Figure 8.3. The vertical stability of disk galaxies (parameterized by the stability parameter Q) as a function of radius. The inner scale length of the massive disks is unstable i.e., the clouds will fall back down the well. The outer disk of massive galaxies and the whole disk of lower mass galaxies is stable, meaning the clouds can remain at higher latitude for much longer timescales. This all presents itself as a dense band of dust clouds mid-plane in the massive disk galaxies and dust clouds scattered in latitude in the low-mass galaxies. Figure reproduced from Dalcanton et al. (2004).

image is a whole spectrum at that pixel's position. Using templates of how starlight is supposed to look, one can then infer the fraction of different types of stars in each pixel or groups of pixels and thanks to the Doppler effect on the absorption lines in stellar atmospheres, the bulk motion of stars in each part of the image. As you can imagine, this is vastly more information on a morphological phenomenon such as a central bar. Figure 8.1 shows what can be done already with a Spitzer Space Telescope image. The inner parts of this galaxy can already be broken into a disk

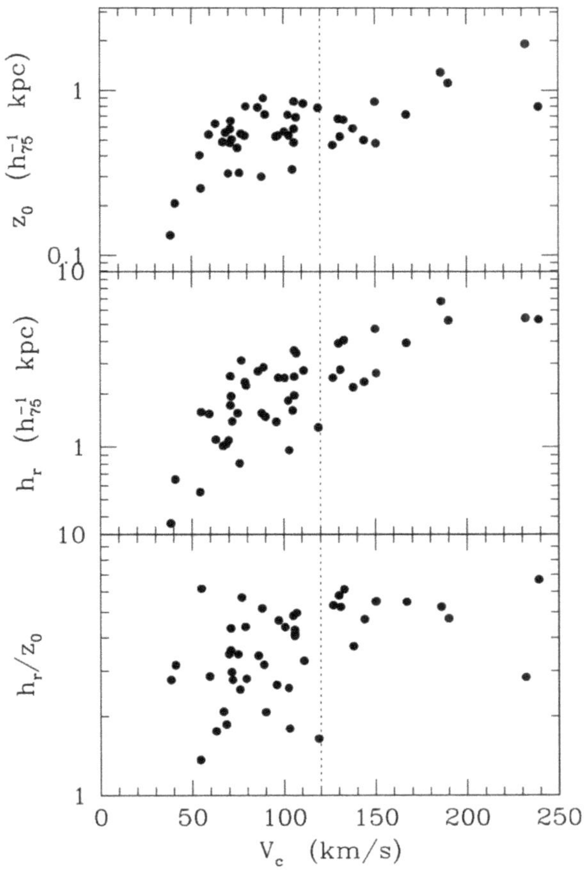

Figure 8.4. *Ks*-band structural parameters derived from two-dimensional fits of edge-on exponential disks as a function of rotation speed for the galaxies (Dalcanton & Bernstein 2000). The top panel shows the vertical scale height (z_0) of the disks and demonstrates that there is substantial overlap in the vertical extent of stellar disks above and below a rotation velocity v = 120 km s^{-1} (maximum speed of the rotation curve). Middle panel: exponential scale lengths (h) of the same disks. Bottom: axial ratio of the disks (h/z_0). Stellar disks with v < 120 km s^{-1} have systematically smaller axial ratios: the more massive ones are more flattened. Figure reproduced from Dalcanton et al. (2004).

and a bulge as well as two bars, a lens, and a central disk. Arguably there seems to be a ring in the residual still! A very complex central region. Yet with the additional MUSE IFU information one can map the velocity along the line of sight (Figure 8.5), down to the dispersion (random motions in the population of stars) and higher-order moment in the velocity distributions along the line of sight.

And finally, because spectral information contains absorption features of the metals[1] and the depth of some features depend on the lifetime of the star, one can

[1] A catch-all term for elements heavier than hydrogen or helium according to astronomy. The periodic table goes hydrogen, deuterium, helium, metals. Oxygen is a metal.

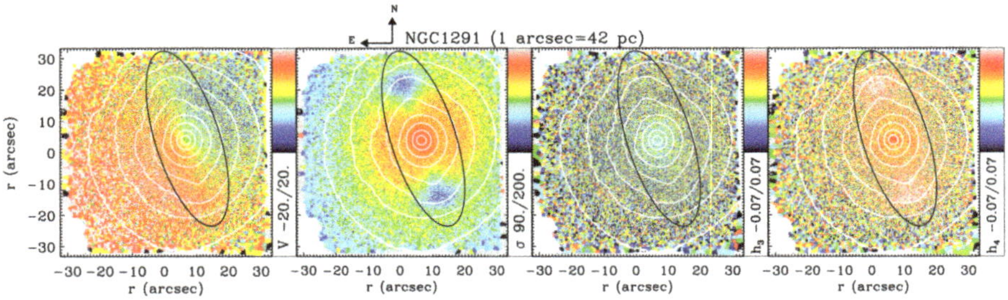

Figure 8.5. 2D distribution of the line-of-sight stellar kinematic moments for NGC1291. From left to right, the subpanels correspond to the stellar velocity V in km s^{-1}, velocity dispersion σ in km s^{-1}, and higher-order Gauss–Hermite moments h_3, and h_4. Figure reproduced from de Lorenzo-Cáceres et al. (2019). © 2019 The Author(s), Published by Oxford University Press on behalf of the Royal Astronomical Society.

now map the age and heavy element content of the populations of stars for each morphological element! Figure 8.6 shows how one can disentangle the stellar ages and metallicities of each component now. How much do they resemble each other? Did the form together (or from the same population) or are they distinct? Is there evidence for mixing? With so much more information on morphological components, there is much more to test models of how certain morphological components form (e.g., how does a bar form and how long does it last?).

8.5 Bar Kinematics

The confluence between morphology and kinematics is when IFU information is available on the bars of galaxies. The primary parameters describing a bar are its radius, $R_{\rm bar}$, relative strength with respect to the disk, $S_{\rm bar}$, pattern speed, $\Omega_{\rm bar}$, and rotation rate R (Cuomo et al. 2020).

The bar pattern speed ($\Omega_{\rm bar}$) is the angular frequency of the bar pattern rotating around it's galaxy center, and it determines the so-called corotation radius $R_{\rm cr}$, which is the radius where the angular velocity of the disk $V_{\rm circ}$ is equal to the speed of the bar ($\Omega_{\rm bar}$). Bars can then be classified according to the rotation rate: $\mathcal{R} = R_{\rm cr}/R_{\rm bar}$ (Elmegreen 1996). Bars are classified according to their inferred rotation rate: bars are fast ($1.0 \leqslant \mathcal{R} \leqslant 1.4$) or slow ($\mathcal{R} > 1.4$).

The radius and strength of the bar are *structural* parameters and can be recovered from the analysis of optical or near-infrared images (Figure 8.1), the pattern speed and rotation rate of the bar are *dynamical* parameters and their determination requires kinematics. Using IFU information on bars, one can infer a bar pattern speed ($\Omega_{\rm bar}$) and the rotation rate of the bar ($\mathcal{R}$).

Now that IFUs with both the spatial and spectroscopic resolution are common on mid-size telescopes for grand surveys and 8 m class telescopes for detailed analysis, the kinematic bar values as well as the structural ones have become available for samples of galaxies (Guo et al. 2019; Garma-Oehmichen et al. 2020; Cuomo et al. 2019, 2021) and much more detail is revealed with MUSE observations (cf the TIMER survey; de Lorenzo-Cáceres et al. 2019; Méndez-Abreu et al. 2020; Neumann et al. 2020; Bittner et al. 2020; Gadotti et al. 2020; Rosado-Belza et al. 2010), linking stellar ages to this secular evolution as well.

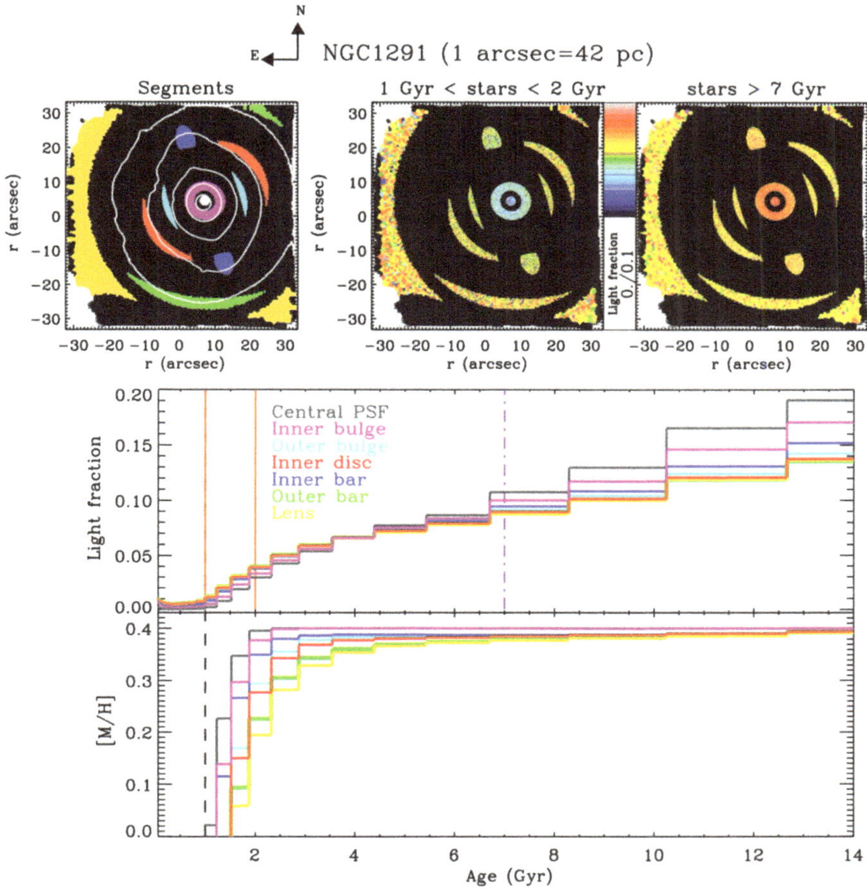

Figure 8.6. The star formation history of the galaxy NGC1291 for which each central morphological element is shown in Figure 8.1 and their velocities in Figure 8.5. The top-left panel identifies where each structural component dominates the observed light. Solid while lines are flux contours The two top-right panels show the 2D distribution of stellar populations within a range of young ages (1–2 Gyr, left panel) and for ages older than 7 Gyr (right panel). The bottom panel shows the contribution of each generation of stars to each component and the distribution of age and metallicities for each morphological component identified in Figure 8.1. Figure reproduced from de Lorenzo-Cáceres et al. (2019). © 2019 The Author(s), Published by Oxford University Press on behalf of the Royal Astronomical Society.

With all this information on bars, one can start exploring the link with dark matter haloes (Collier 2020; Collier & Madigan 2021) and other morphological features (Buta et al. 2009).

8.6 Rings: Resonances

Once there is a pattern—such as a bar—that deviates from the radially uniform disk, one can expect resonance patterns to emerge, provided the perturber (e.g., bar) is long-lived. Like the moons of Jupiter, the stars in the disk will start to organize into orbits that are in resonance. The best examples are inner and outer Lindblad resonances. Either the star resonating is being overtaking by the perturber (outer) or it is co-rotating with the perturber, or it is overtaking the perturber. These inner and outer Lindblad resonance rings (ILR and OLR respectively) is how we think how many rings in disks start (recall this from Chapter 4). But now that we have a rotation curve from radio observations and for example bar kinematic and strength parameters, one can directly compare the predicted kinematics to the observations!

8.7 Spiral Arms

Spiral arms similarly could be started as a resonance effect. The start of the spiral pattern could be a fly-by (e.g., M51) or the formation of a bar (Mondal & Chattopadhyay 2021).

8.8 Tidal Features

The tidal features we encountered in Chapter 7 are the result of a satellite galaxy falling into the potential of a larger galaxy. Satellite galaxies are stretched into streams by a process called "phase wrapping." The clump of mostly stars (point particles) that is the satellite galaxy has a side that falls into the potential first and a side that falls in last. The near side gains speed (kinetic energy) with respect to the far side earlier, speeding away from the center and far side. This tidal effect keeps stretching the dwarf galaxy until it is so diffuse it only shows as a LSB stream.

With ever deeper imaging observations (Martinez-Delgado et al. 2021), such events are found to be common. The morphology of the stream depends on both the angle of incidence (does it hit the main galaxy head-on or come in at an angle). Figure 8.7 shows three scenarios from kinematic models and matches them with what can be seen in DES imaging.

With enough examples of streams soon available in the "nearby" (<150 Mpc), we can match scenarios and study incident statistics. How often have head-on collisions happened? How many collisions happen each Gyr? With the Vera Rubin Observatory starting operations soon, much more can be expected in this area.s

8.9 Ellipticals versus Disks

Figure 8.8 illustrates a better way to think about elliptical galaxies rather than the Hubble tuning fork we encountered in Chapter 1. Rather than just thinking in terms of "early type galaxies"[2] that are defined by morphological absences (no spiral arms or clear dust lanes), to also think of ellipticals in terms of their kinematics, the measure of rotation.

The two component Sérsic fits of elliptical (red) galaxies can now be linked to their level of rotation and the relative prominence of disks in the profile. Figure 8.9 illustrates how galaxies are no longer just classified using morphology but morpho-kinematics. It is now possible to classify spheroidal (or "early type galaxies") according to their total angular momentum (Emsellem et al. 2007; Cappellari et al. 2007).

[2] I still do not like this term but terminology persists.

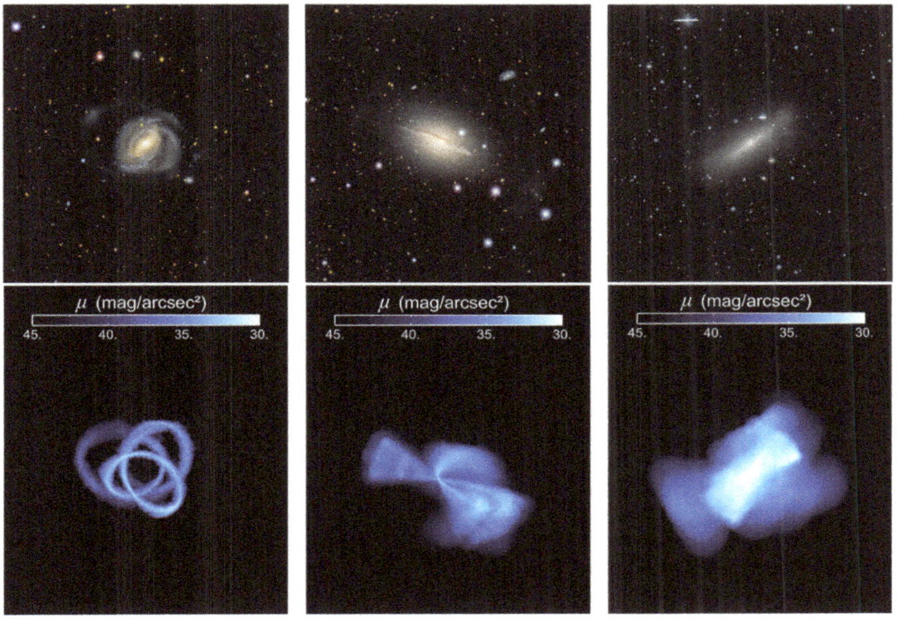

Figure 8.7. The external perspective snapshots for the three morphological types classified by their origin (Johnston et al. 2008): "great circle" streams (left panels) originated from satellites accreted 6–10 Gyr ago on nearly circular orbits; "shells" or "umbrellas" (middle panels) arise from accretion events within the past ~8 Gyr with radial orbits; and "mixed"-type tidal remnants (right panels) formed from ancient (more than 10 Gyr ago) accretion events that have fully mixed along their orbits. The top panels are image cutouts obtained from the DESI Legacy Imaging Surveys data matching each morphological case of tidal stream. The great circle stream is NGC 577 (top left). The shell-like features are around NGC 681 (middle). The "mixed" remnant around NGC 5866 (top-right panel). Figure reproduced with permission from Carlin et al. (2016) (Copyright © 2016, Springer International Publishing Switzerland) and Martinez-Delgado et al. (2021).

8.10 Angular Momentum

In the previous section, we already encountered total angular momentum (λ) for the whole galaxy. The total angular momentum of a galaxy is the vector sum of every element of mass, its velocity in its orbit:

$$\vec{J} = \Sigma \vec{v} \times m \times r. \tag{8.1}$$

This galaxy-wide kinematic measure depends on a lot of different measurements: position in the galaxy, estimates of kinematics (rotation and random motion) as well as mass-to-light ratio to estimate the mass in each element of the galaxy since we trace mass indirectly by the light. Disk galaxies have a much higher total angular momentum than the spheroidals (Fall 1983; Romanowsky & Fall 2012). The organized motion of a disk rotating around its center means the angular momentum vectors of all the elements compound. To be able to compare very different galaxies, the specific angular momentum j is used:

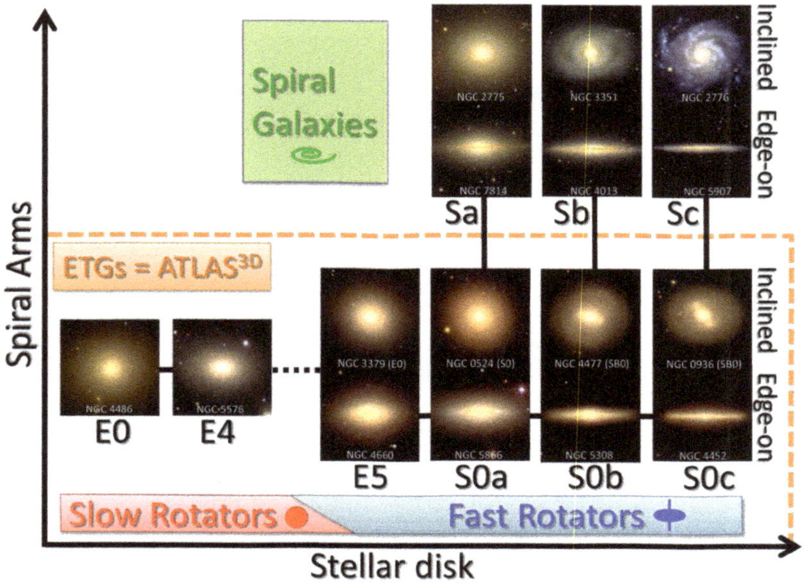

Figure 8.8. The morphological-kinematic classification (Cappellari et al. 2011a) indicating their target sample of "early type galaxies (ETG)" targeted by ATLAS^{3D} including fast-rotators ETGs and slow rotators ETGs. The edge-on fast rotators appear morphologically equivalent to S0s, or to flat ellipticals with disky isophotes. Many of the apparently-round fast-rotators display bars or dusty disks, indicating that they are far from edge-on. Figure adapted from Cappellari et al. (2011a). © 2011 The Authors, Monthly Notices of the Royal Astronomical Society © 2011 RAS.

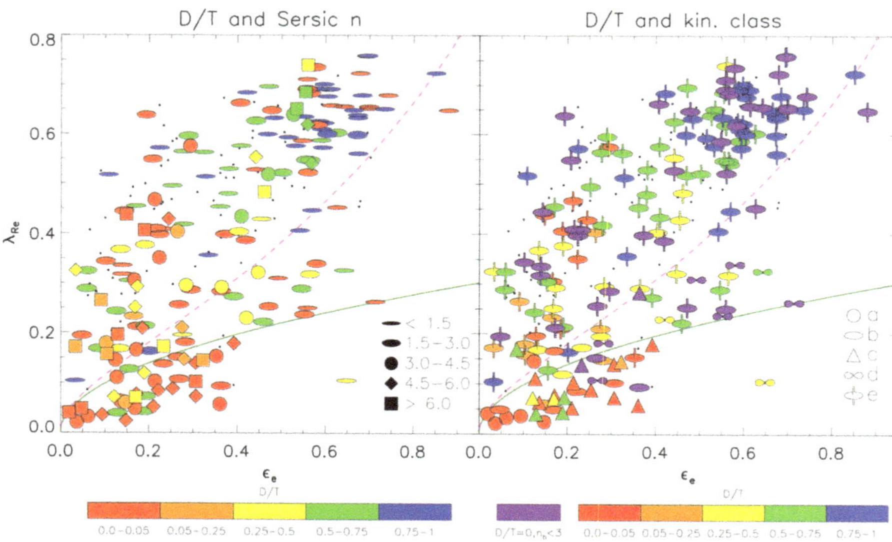

Figure 8.9. The angular momentum (λ_R) as a function of galaxy ellipticity (ϵ) for the galaxies in the ATLAS^{3D} survey. On the left the Sérsic fit results with symbols showing the Sérsic index (n) and color-coding the disk-to-total light ratio (D/T). Total angular momentum increases with a greater disk component. Right shows the same but with the kind of rotator shown with the symbols: a—non rotating galaxies, b—featureless non-regular rotators, c—KDC, d—2σ, and e—regular rotators. Figure adapted from Krajnović et al. (2013a). © 2012 The Authors, Published by Oxford University Press on behalf of the Royal Astronomical Society.

$$j = J/M \tag{8.2}$$

where M is the total mass of all the elements combined. As we saw above, this quantity is already very useful to classify spheroidal galaxies in a meaningful way, i.e., physical differences rather than viewing angle.

The next question is if classification with angular momentum can be applied to disk galaxies and what that tells us about their history and nature. Figure 8.10 shows the relationship between the specific angular momentum (j_b), bulge-to-total (β), and mass (M_b) in 16 nearby spiral galaxies (Obreschkow & Glazebrook 2014). In spiral disk galaxies, a lot of the angular momentum and mass are not in stars but in the gas of the disk. It is therefore important to account for both gas and stars, i.e., all the Baryonic mass. Figure 8.10 shows how the relative importance of the spheroidal component (classical bulge) affects the specific angular momentum–mass relation for disk galaxies.

The mass-angular momentum relation for disk galaxies is a possible explanation why disks follow an exponential profile so neatly (Herpich et al. 2017). Disks start out with turbulence in the process of formation (Obreschkow et al. 2015) and yet

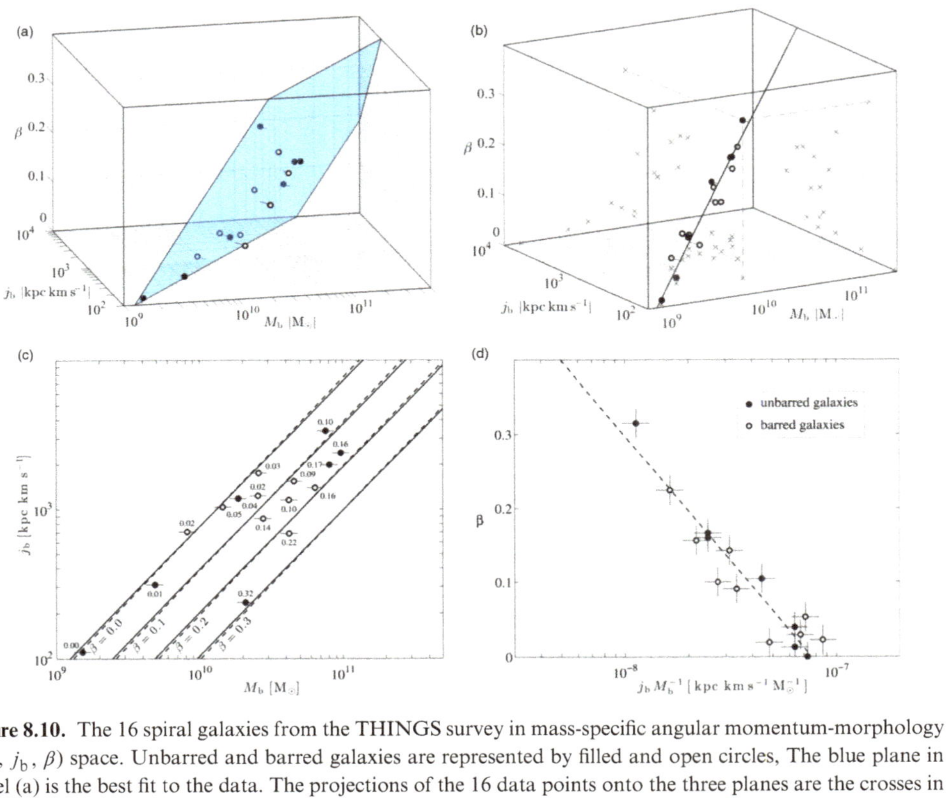

Figure 8.10. The 16 spiral galaxies from the THINGS survey in mass-specific angular momentum-morphology (M_b, j_b, β) space. Unbarred and barred galaxies are represented by filled and open circles, The blue plane in panel (a) is the best fit to the data. The projections of the 16 data points onto the three planes are the crosses in panel (b). The blue plane is a solid line in panel (b). Discrete values of the bulge-to-total (β) are solid lines in panel (c). The single dashed line in panel (d) is the best fit between β and $\log_{10}(j_b/M_b)$. Figure reproduced from Obreschkow & Glazebrook (2014).

organize themselves into neat exponentials. Tracking the evolution of angular momentum of galaxies in both simulations (Contreras et al. 2017; Lagos et al. 2017) and observations (Obreschkow et al. 2016; Murugeshan et al. 2019, 2020) can now be a way to understand galaxy population's morphological and kinematic evolution.

8.11 Assignments

There three assignments to go with kinematics and galaxies: "Rotation Curve" (Jupyter notebook 12), "The Tully Fisher Relation" (Jupyter notebook 13), and "The Fundamental Plane of Elliptical Galaxies" (Jupyter notebook 14).

8.12 Further Reading

The study of morpho-kinematic classification originates for a large part in the structural study of nearby spheroidal galaxies in two projects: the SAURON (Bacon et al. 2001; de Zeeuw et al. 2002; Emsellem et al. 2004; Cappellari et al. 2006, 2007; Sarzi et al. 2006; Kuntschner et al. 2006; Falcón-Barroso et al. 2006; McDermid et al. 2006; Emsellem et al. 2007; Peletier et al. 2007; Krajnović et al. 2008; Jeong et al. 2009; Scott et al. 2009; Shapiro et al. 2010; Sarzi et al. 2010) and the ATLAS^{3D} (Cappellari et al. 2011a, 2011b, 2013a, 2013b; Krajnović et al. 2011, 2013a, 2013b; Emsellem et al. 2011; Young et al. 2011, 2014; Bois et al. 2011; Khochfar et al. 2011; Duc et al. 2011, 2015; Davis et al. 2011a, 2011b, 2013, 2014; Crocker et al. 2012; Lablanche et al. 2012; Serra et al. 2012, 2014; Bayet et al. 2013; Alatalo et al. 2013; Sarzi et al. 2013; Scott et al. 2013; Martig et al. 2013; Weijmans et al. 2014; Naab et al. 2014; McDermid et al. 2015; Nyland et al. 2016) projects. These studied both the kinematics as well as star formation, gas content etc. for large samples of elliptical (ETG) galaxies.

For the use of angular momentum in disk galaxies, the Obreschkow (Obreschkow & Glazebrook 2014) is a good start.

References

Alatalo, K., Davis, T. A., Bureau, M., et al. 2013, MNRAS, 432, 1796
Bacon, R., Copin, Y., Monnet, G., et al. 2001, MNRAS, 326, 23
Bayet, E., Bureau, M., Davis, T. A., et al. 2013, MNRAS, 432, 1742
Bittner, A., Sánchez-Blázquez, P., Gadotti, D. A., et al. 2020, A&A, 643, A65
Bois, M., Emsellem, E., Bournaud, F., et al. 2011, MNRAS, 416, 1654
Bosma, A., van der Julst, J. M., Sullivan, W. T. III, et al. 1977, A&A, 57, 373
Buta, R. J., Knapen, J. H., Elmegreen, B. G., et al. 2009, AJ, 137, 4487
Cappellari, M., Bacon, R., Bureau, M., et al. 2006, MNRAS, 366, 1126
Cappellari, M., Emsellem, E., Bacon, R., et al. 2007, MNRAS, 379, 418
Cappellari, M., Emsellem, E., Krajnovic, D., et al. 2011a, MNRAS, 413, 813
Cappellari, M., Emsellem, E., Krajnovic, D., et al. 2011b, MNRAS, 416, 1680
Cappellari, M., McDermid, R. M., Alatalo, K., et al. 2013a, MNRAS, 432, 1862
Cappellari, M., Scott, N., Alatalo, K., et al. 2013b, MNRAS, 432, 1709

Carlin, J. L., Beaton, R. L., Martinez-Delgado, D., et al. 2016, in Astrophysics and Space Science Library, Vol. 420, Tidal Streams in the Local Group and Beyond (Cham: Springer), 219
Collier, A. 2020, MNRAS, 492, 2241
Collier, A., & Madigan, A.-M. 2021, ApJ, 915, 23
Contreras, S., Padilla, N., & Lagos, C. D. P. 2017, MNRAS, 472, 4992
Crocker, A., Krips, M., Bureau, M., et al. 2012, MNRAS, 421, 1298
Cuomo, V., Aguerri, J. A. L., Corsini, E. M., & Debattista, V. P. 2020, A&A, 641, A111
Cuomo, V., Lee, Y. H., Buttitta, C., et al. 2021, A&A, 649, A30
Cuomo, V., Lopez Aguerri, J. A., Corsini, E. M., et al. 2019, A&A, 632, A51
Dalcanton, J. J., & Bernstein, R. A. 2000, in ASP Conf. Ser. 197, Dynamics of Galaxies: From the Early Universe to the Present, ed. F. Combes, G. A. Mamon, & V. Charmandaris, 161
Dalcanton, J. J., Yoachim, P., & Bernstein, R. A. 2004, ApJ, 608, 189
Davis, T. A., Alatalo, K., Bureau, M., et al. 2013, MNRAS, 429, 534
Davis, T. A., Alatalo, K., Sarzi, M., et al. 2011a, MNRAS, 417, 882
Davis, T. A., Bureau, M., Young, L. M., et al. 2011b, MNRAS, 414, 968
Davis, T. A., Young, L. M., Crocker, A. F., et al. 2014, MNRAS, 444, 3427
de Lorenzo-Cáceres, A., Sanchez-Blazquez, P., Mendez-Abreu, J., et al. 2019, MNRAS, 484, 5296
de Zeeuw, P. T., Bureau, M., Emsellem, E., et al. 2002, MNRAS, 329, 513
Duc, P.-A., Cuillandre, J.-C., Karabal, E., et al. 2015, MNRAS, 446, 120
Duc, P.-A., Cuillandre, J.-C., Serra, P., et al. 2011, MNRAS, 417, 863
Elmegreen, B. 1996, in ASP Conf. Ser. 91, IAU Colloq. 157: Barred Galaxies, ed. R. Buta, D. A. Crocker, & B. G. Elmegreen (San Francisco, CA: ASP), 197
Emsellem, E., Cappellari, M., Krajnovic, D., et al. 2011, MNRAS, 414, 888
Emsellem, E., Cappellari, M., Krajnovic, D., et al. 2007, MNRAS, 379, 401
Emsellem, E., Cappellari, M., Peletier, R. F., et al. 2004, MNRAS, 352, 721
Falcón-Barroso, J., Bacon, R., Bureau, M., et al. 2006, MNRAS, 369, 529
Fall, S. M. 1983, in IAU Symp. 100, Internal Kinematics and Dynamics of Galaxies, ed. E. Athanassoula, 391
Gadotti, D. A., Bittner, A., Falcón-Barroso, J., et al. 2020, A&A, 643, A14
Garma-Oehmichen, L., et al. 2020, MNRAS, 491, 3655
Guo, R., Mao, S., Athanassoula, E., et al. 2019, MNRAS, 482, 1733
Herpich, J., Tremaine, S., & Rix, H.-W. 2017, MNRAS, 467, 5022
Holwerda, B. W., Dalcanton, J. J., Radburn-Smith, D., et al. 2012, ApJ, 753, 25
Holwerda, B. W., Kelvin, L., Baldry, I., et al. 2019, AJ, 158, 103
Jeong, H., Yi, S. K., Bureau, M., et al. 2009, MNRAS, 398, 2028
Johnston, S., Taylor, R., Bailes, M., et al. 2008, ExA, 22, 151
Khochfar, S., Emsellem, E., Serra, P., et al. 2011, MNRAS, 417, 845
Krajnović, D., Alatalo, K., Bois, L., et al. 2013a, MNRAS, 432, 1768
Krajnović, D., Bacon, R., Cappellari, M., et al. 2008, MNRAS, 390, 93
Krajnović, D., Emsellem, E., Cappellari, M., et al. 2011, MNRAS, 414, 2923
Krajnović, D., Karick, A. M., Davies, R. L., et al. 2013b, MNRAS, 433, 2812
Kuntschner, H., Emsellem, E., Bacon, R., et al. 2006, MNRAS, 369, 497
Lablanche, P.-Y., Cappellari, M., Emsellem, E., et al. 2012, MNRAS, 424, 1495
Lagos, C. D. P., Theuns, T., Stevens, A. R. H., et al. 2017, MNRAS, 464, 3850
Martig, M., Crocker, A. F., Bournaud, F., et al. 2013, MNRAS, 432, 1914
Martinez-Delgado, D., Cooper, A. P., Roman, J., et al. 2021, arXiv:2104.06071

McDermid, R. M., Alatalo, K., Blitz, L., et al. 2015, MNRAS, 448, 3484
McDermid, R. M., Emsellem, E., Shapiro, K. L., et al. 2006, MNRAS, 373, 906
Méndez-Abreu, J. de Lorenzo-Cáceres A., & Timer Team 2020, in IAU Proc. 353, Galactic Dynamics in the Era of Large Surveys, 168
Mondal, D., & Chattopadhyay, T. 2021, CeMDA, 133, 43
Murugeshan, C., Kilborn, V., Jarrett, T., et al. 2020, MNRAS, 496, 2516
Murugeshan, C., Kilborn, V., Obreschkow, D., et al. 2019, MNRAS, 483, 2398
Naab, T., Oser, L., Emsellem, E., et al. 2014, MNRAS, 444, 3357
Neumann, J., Fragkoudi, F., Pérez, I., et al. 2020, A&A, 637, A56
Nyland, K., Young, L. M., Wrobel, J. M., et al. 2016, MNRAS, 458, 2221
Obreschkow, D., & Glazebrook, K. 2014, ApJ, 784, 26
Obreschkow, D., Glazebrook, K., Bassett, R., et al. 2015, ApJ, 815, 97
Obreschkow, D., Glazebrook, K., Kilborn, V., & Lutz, K. 2016, ApJL, 824, L26
Peletier, R. F., Falcon-Barroso, J., Bacon, R., et al. 2007, MNRAS, 379, 445
Romanowsky, A. J., & Fall, S. M. 2012, ApJS, 203, 17
Rosado-Belza, D., Falcón-Barroso, J., Knapen, J. H., et al. 2020, A&A, 644, A116
Rubin, V. C. 1980, Mercu, 9, 78
Sarzi, M., Alatalo, K., Blitz, L., et al. 2013, MNRAS, 432, 1845
Sarzi, M., Falcon-Barroso, J., Davies, R. L., et al. 2006, MNRAS, 366, 1151
Sarzi, M., Shields, J. C., Schawinski, K., et al. 2010, MNRAS, 402, 2187
Scott, N., Cappellari, M., Davies, R. L., et al. 2009, MNRAS, 398, 1835
Scott, N., Cappellari, M., Davies, R. L., et al. 2013, MNRAS, 432, 1894
Serra, P., Oosterloo, T., Morganti, R., et al. 2012, MNRAS, 422, 1835
Serra, P., Oser, L., Krajnovic, D., et al. 2014, MNRAS, 444, 3388
Shapiro, K. L., Falcon-Barroso, J., van de Ven, G., et al. 2010, MNRAS, 402, 2140
Weijmans, A.-M., de Zeeuw, P. T., Emsellem, E., et al. 2014, MNRAS, 444, 3340
Young, L. M., Bureau, M., Davis, T. A., et al. 2011, MNRAS, 414, 940
Young, L. M., Scott, N., Serra, P., et al. 2014, MNRAS, 444, 3408

Chapter 9

Wavelength Dependence

Galaxy appearance changes with wavelength as different physical processes dominate the spectrum. In the ultraviolet and blue side of optical, it is the young, massive stars, in the red side of optical and near-infrared it is dominated by the older populations of stars. In the mid-infrared there are families of carbon molecules known as polycyclic aromatic hydrocarbons.[1] Beginning in the mid-infrared interstellar grains, the size of a few micron and larger dominate the blackbody radiation. Finally in the radio, the central supermassive black hole's activity makes itself known as well as stellar remnants and many other processes. And at every wavelength the galaxy appears differently as each component is distributed differently throughout.

9.1 Radio Galaxies

Radio continuum can be reliably detected with even a small telescope. With ever larger (synthesis) radio telescopes, higher spatial resolution resolved radio sources into very distinct shapes and morphologies. A central engine for radio galaxies is the active galactic nucleus (AGN); the supermassive black hole in the center of the galaxy whose infall of gas converts some of that material into two jets moving at relativistic speeds (Figures 9.1 and 9.2). If one looks straight down the jet coming off the supermassive black hole, this is a quasar.

With two jets, one beaming emission toward the observer and one away or both mostly perpendicular to the observer (in the plane of the sky) two types of radio galaxies can distinguished: one where we can see both jets (looking sideways at both), Fanaroff–Riley Class I (FR-I), and the brightness diminishes along the jet.

[1] Your nose would know them mostly as "diesel exhaust" but it's stellar exhaust here.

doi:10.1088/2514-3433/ac2c7dch9

Figure 9.1. A color composite image of Centaurus A, one of the best-known radio galaxies in the sky in sub-mm, X-ray, and optical data. The image reveals the radio lobes and jets emanating from the active galaxy's central black hole. This is a composite of images obtained with three instruments, operating at very different wavelengths. The 870 micron submillimeter data, from LABOCA on APEX, are shown in orange. X-ray data from the Chandra X-ray Observatory are shown in blue. Visible light data from the Wide Field Imager (WFI) on the MPG/ESO 2.2 m telescope located at La Silla, Chile, show the stars and the galaxy's characteristic dust lane in close to "true color." Credit: ESO/WFI (Optical); MPIfR/ESO/APEX/A.Weiss et al. (Submillimetre); NASA/CXC/CfA/R.Kraft et al. (X-ray), from https://www.eso.org/public/images/eso0903a/.

And there are those radio where we view closer along the axis of the jet are classified as Fanaroff–Riley Class II (FR-II)[2] and these exhibit increasing brightness in the outer lobes.

Some examples of both these classes of radio galaxy are shown in Figure 9.3. With the new and refurbished radio telescopes now in operation surveying the sky,

[2] Once again we astronomers gave classes a number and not a descriptive name.

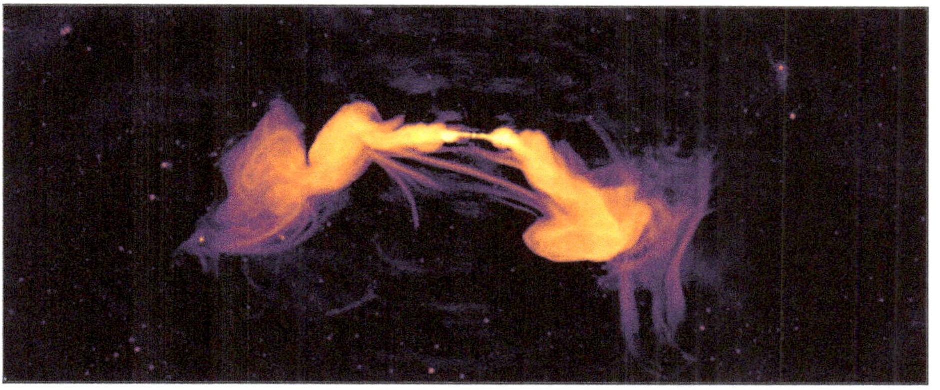

Figure 9.2. ESO 137-006 is a galaxy in the Norma cluster of galaxies, and one of the brightest objects in the Southern sky at radio wavelengths. It is characterized by two very bright and bent lobes of radio emission formed by the active galactic nucleus (AGN) of ESO 137-006, which hosts a growing supermassive black hole at its center. These new features have been uncovered in this galaxy in the form of multiple, extremely collimated threads of radio emission connecting the lobes of the galaxy. Image by Ramatsoku et al. (2020). Reprinted with permission from ASTRON, NOW.

we can expect thousands if not millions of both classes to be found. Not only that, the inner structure of the jets in these galaxies are now exposed in great detail (Figures 9.2 and 9.3).

9.2 HI Maps

Also in the radio is the 21 cm emission line by hydrogen. Because hydrogen is so ubiquitous and the line does not need a lot to be excited (it is due to the flip in spin by atomic hydrogen's single electron), the 21 cm emission maps the presence of atomic hydrogen throughout a galaxy (Figure 9.4). And because hydrogen is the principle component in making new stars as well as a fantastic kinematic tracer, radio astronomers have done their best to map HI in nearby galaxies. HI appearance is completely different from the stellar one in most galaxies. No more bulge and not particularly exponential disks. The range in values in HI surface density is substantially less than in stellar light where the dynamic range spans orders of magnitude.

HI disks also extend much further out than the stellar disk. So far out in fact that some started talking about "pristine," i.e., uncontaminated by stars or any other elements. That is unlikely (we find metals even further out still). Figure 9.5 shows very nicely how much further out the HI is detected compared to the stars.

And it is unlikely to stop there. Figure 9.6 shows two integration depths for HI observations: we suspect that, like with low surface brightness (LSB) features, there are low column density features of gas far around galaxies yet to be discovered. Initial surveys such as HALOGAS (Heald et al. 2011a, 2011b; Zschaechner et al. 2011b, 2012; Gentile et al. 2013; de Blok et al. 2014; Heald 2015; Ianjamasimanana et al. 2018; Pingel et al. 2018; Richter et al. 2018; Marasco et al. 2019; Reynolds et al. 2020) already show there is much more to the gas in disk galaxies than we have seen so far. With the new MeerKAT radio telescope in South Africa, we can expect even more

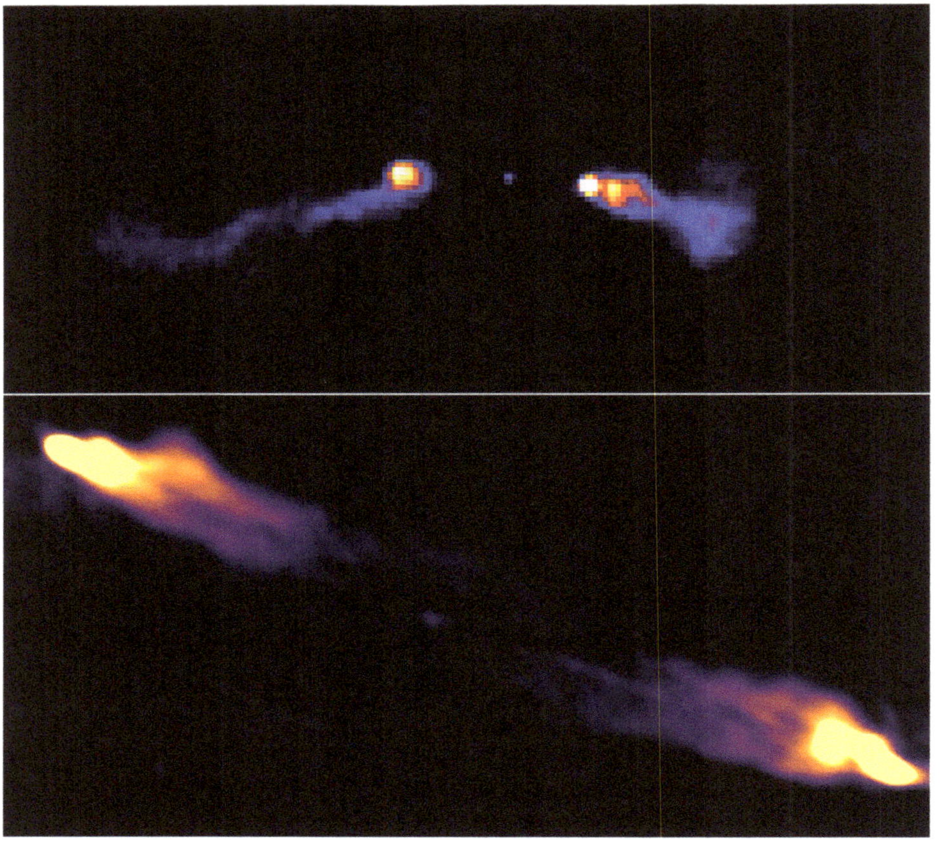

Figure 9.3. Fanaroff–Riley 1 and 2 (FR-1 and FR-2, top and bottom panel respectively) labels come from a classification system developed by B. L. Fanaroff and J. M. Riley that is used to identify radio galaxies with AGN based on the brightness of their radio emissions. FR-1's are radio sources whose luminosity decreases as the distance from the central galaxy host increases and can be signposts for galaxy clusters. FR-II's are sources that exhibit increasing brightness in the outer lobes. Credit: NRAO/AUI/NSF, from https://public.nrao.edu/gallery/radio-galaxy/.

of these revelations, showing shoals of gas further out from what we used to think of as the galaxy.

However, we can describe the atomic gas of galaxies in similar ways as stellar ones: there are signs of tidal interactions (Hibbard et al. 2001), spiral structure and one can define morphometrics on such images (Holwerda et al. 2011b, 2011c, 2011d, 2011e, 2011f; Giese et al. 2016; Reynolds et al. 2020). But because HI observations typically lag a little behind optical ones (historical reasons and HI is catching up in sensitivity and resolution), HI morphological properties are often defined with respect to optical ones (e.g., spiral arms). And while there is certainly an interplay, we must remember that this gas we are tracing (i.e., something that has an internal pressure) not a cloud of particles only governed by gravity.

Figure 9.4. The HI in Nearby Galaxies Survey (THINGS) with every galaxy shown to scale. It shows the wide variety of atomic hydrogen morphology in spiral and irregular disk galaxies. M101 is visible bottom left as one of the favorites for observing. A to-scale image of the Milky Way and an artist impression are in the center. Image reprinted with permission from https://www2.mpia-hd.mpg.de/THINGS/THINGS_Poster.html. See also Walter et al. (2008).

For example, a topic that is almost exclusively studied in HI is holes: gaps in the fairly uniform HI disk. And because HI information by definition has a velocity component, one can study the holes in disks as 3D structures.

Figure 9.7 shows an example from M101 (Kamphuis et al. 1991) where not only is there a clear hole in the disk, there is evidence for a chimney of gas that is blown out as a result of the energetic event that created it (Figure 9.8).

9.3 Sub-mm and Mid-infrared

The next wavelength is sub-mm to mid-infrared, typically expressed in micron (10 to 1000 micron). This wavelength range is dominated by the thermal emission of interstellar dust grains. Figure 9.16 shows 350 micron (μm) images from the Herschel Space Telescope.

The challenge in the sub-mm and mid-infrared morphologies are that spatial resolution only became recently available with the Herschel Space Telescope and the Atacama Millimeter Array and the interpretation depends on how much the dust grains are being illuminated by the surrounding stars. Unlike the stellar emission or the radio, dust emission is the convolution of both density of dust and the amount of irradiation.

Figure 9.5. A composite of the stellar light (yellow) and the HI data (blue) of galaxy NGC 6946 (the "fireworks galaxy" thanks to all its star formation and supernovae). The HI extends to well beyond isophotal radii of the starlight. Credit: Netherlands Institute for Radio Astronomy Board Design Gijs (slidetodoc.com).

The Herschel Space Telescope was such a gamechanger for sub-mm and the morphology of dust in galaxies because for the first time it resolved it into morphological features in nearby galaxies (see for examples Figures 4.21, 9.9, 9.13, and 9.16). These galaxies were no longer unresolved single blobs of emission.[3]

What the improved resolution bought astronomy in terms of information is illustrated in Figure 9.9. Here the disk of a small, nearby edge-on galaxy is resolved into discrete blobs (clouds of dust and gas) and can be directly compared to the disposition of the atomic hydrogen (HI).

[3] Herschel's forerunner, the IRAS satellite did not resolve galaxies and as such did "blobology" my favorite non-word in astronomy.

Figure 9.6. Comparison of shallow THINGS (Walter et al. 2008) and deep HALOGAS (Heald et al. 2011a) HI observations of NGC 925 at the same spatial resolution. Left: the HI distribution of NGC 925 from THINGS, convolved to the Westerborg Synthesis Radio Telescope (WSRT) resolution. Lowest contour at 9×10^{19} cm^{-2}, with each contour double the previous value. Center: false-color image of the baryonic components: blue shows the HI gas distribution derived from THINGS, cyan the distribution of young stars derived from GALEX ultraviolet data. Right: the HI distribution as observed by the deep WSRT HALOGAS survey. Lowest contour 1.8×10^{19} cm^{-2} with each contour double the previous value. Figure reprinted with permission from de Blok et al. (2016).

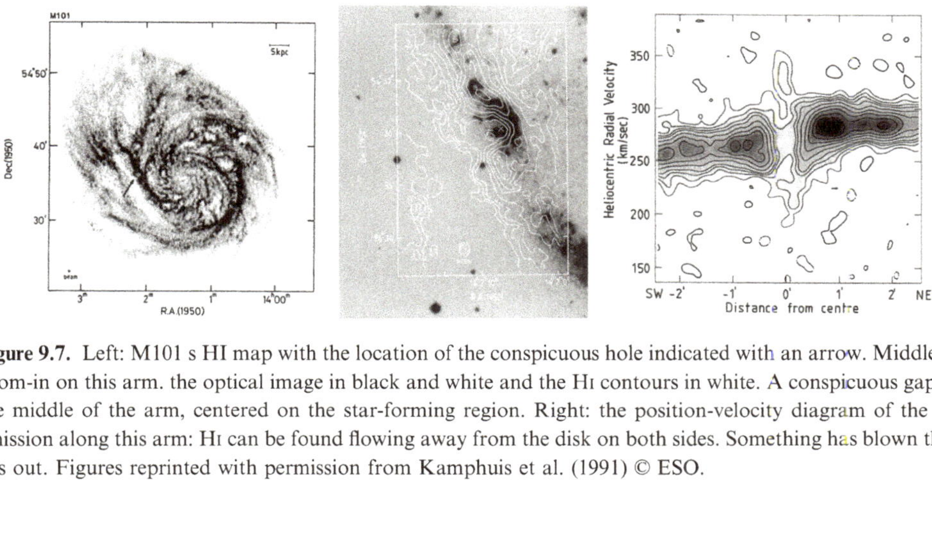

Figure 9.7. Left: M101 s HI map with the location of the conspicuous hole indicated with an arrow. Middle: a zoom-in on this arm. the optical image in black and white and the HI contours in white. A conspicuous gap in the middle of the arm, centered on the star-forming region. Right: the position-velocity diagram of the HI emission along this arm: HI can be found flowing away from the disk on both sides. Something has blown that gas out. Figures reprinted with permission from Kamphuis et al. (1991) © ESO.

9.3.1 PAWS

Combined with dust thermal emission, one can use the CO lines in the sub-mm to trace denser molecular (hydrogen and other molecules) in nearby galaxies. An example of such a result is the CO(1-0) observations from the PdBI Arcsecond Whirlpool Survey (PAWS), Schinnerer et al. (2013) tracing the bulk molecular gas in conjunction with ancillary information on all the other interstellar medium (ISM) phases and star formation. In the Northern Hemisphere, sub-mm observations are best obtained with the Plateau de Bure interferometer (PdBI; Meidt et al. 2013;

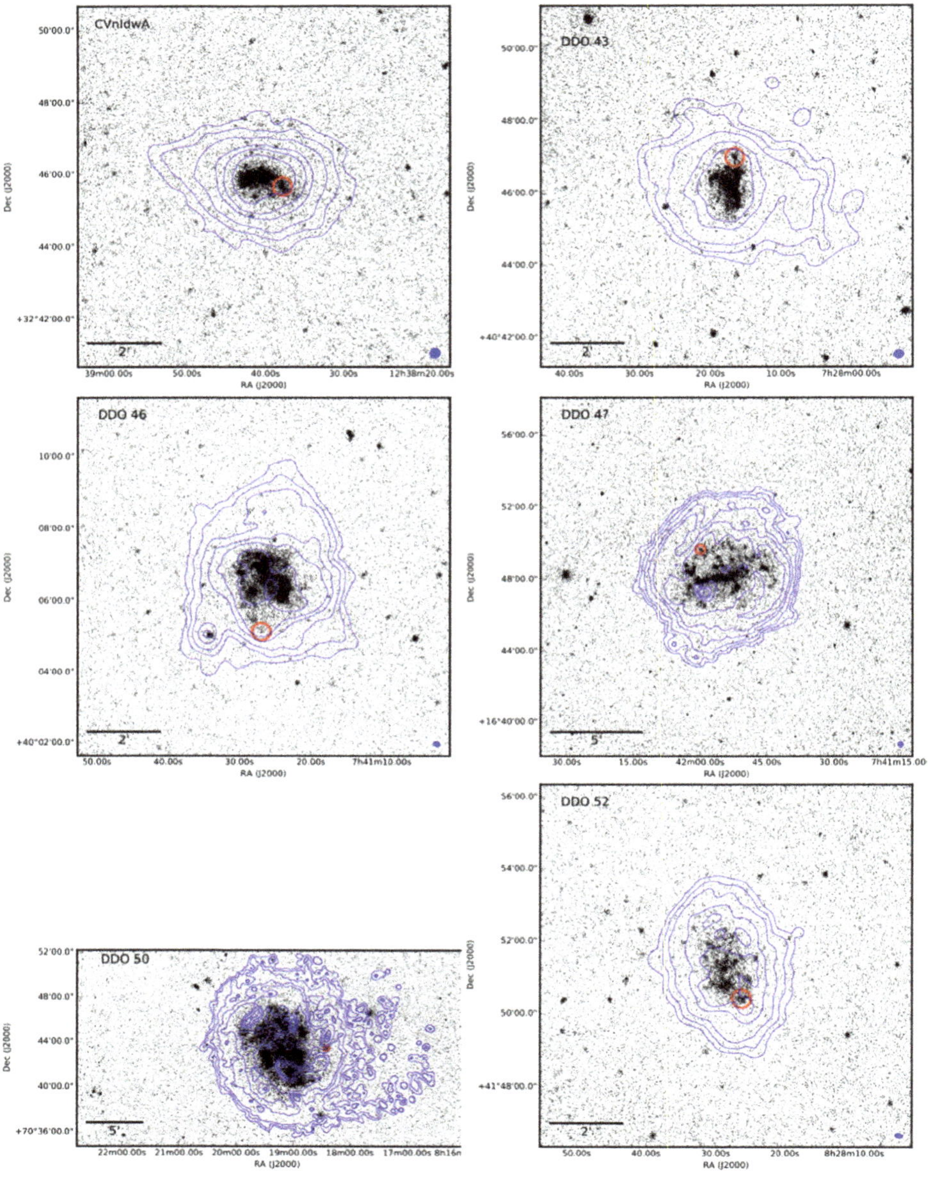

Figure 9.8. The far-ultraviolet grayscale images of the Little-THINGS galaxies with blue HI contours overlaid. The red circle outlines the distinct FUV region at the furthest radius that still belongs to the galaxy. The blue contours are for column densities of 5, 30, 100, 300, 500, 1000, and 3000 × 10^{18} atoms per cm^2 from Hunter et al. (2012). The blue ellipse in the bottom right of each galaxy panel is the HI data's resolution (beam size). Figure reproduced from Hunter et al. (2016).

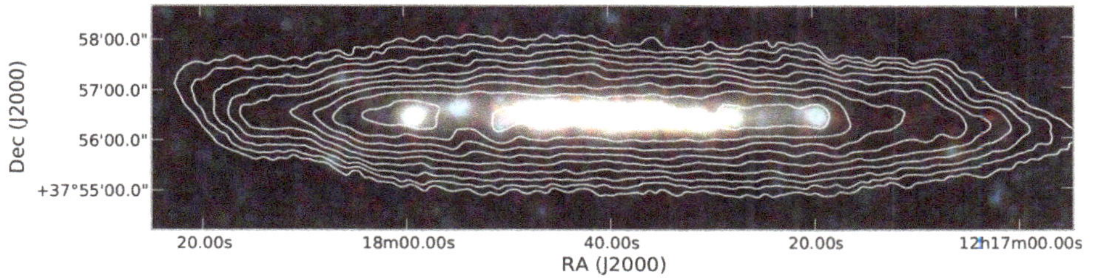

Figure 9.9. A color composite of the SPIRE maps of NGC 4244, 250 (red), 350 (green), and 500 μm (blue) with the HI contours from Zschaechner et al. (2011a) contours at 6.4×10^{19} and factors of 2. The SPIRE flux stays within the 8.1×10^{19} cm^{-2}, the second to highest contour. Figure reprinted with permission from Holwerda et al. (2011) © ESO.

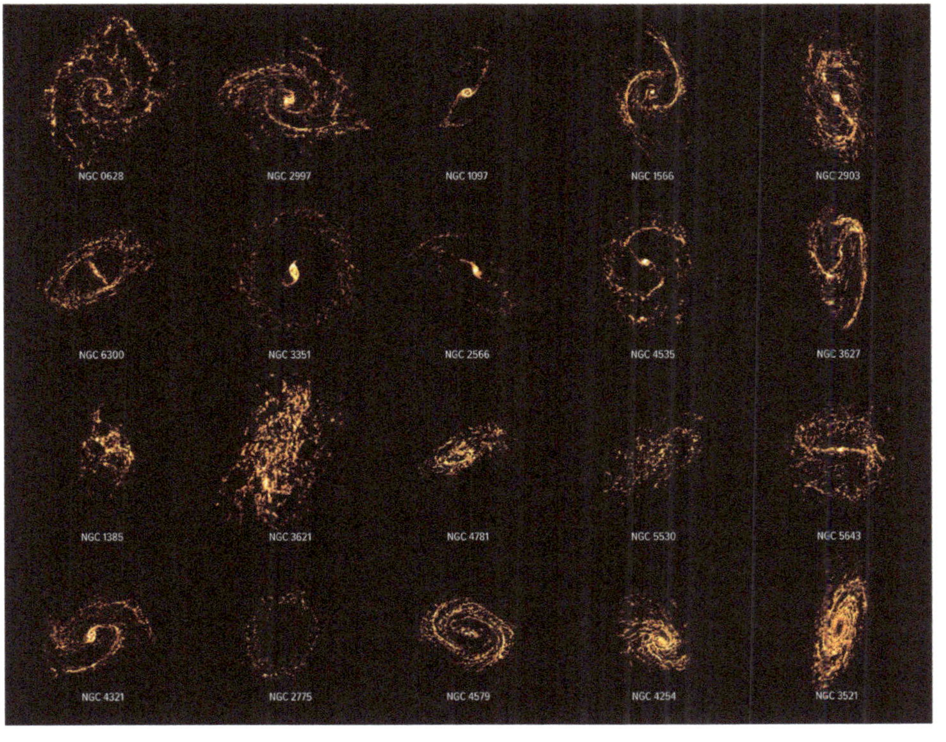

Figure 9.10. ALMA maps of nearby galaxies targeted by the Physics at High Angular Resolution in Nearby GalaxieS project. Star formation and stellar populations are studied in detail with Hubble Space Telescope observations. The ALMA observations find the denser clouds of dust and gas from which stars are formed in these galaxies. The environment in every galaxy or every stellar nursery spans a range of physical states. Figure from https://public.nrao.edu/news/alma-phangs-stellar-nursery-census/ accompanying (Leroy et al. 2021). Credit: ALMA (ESO/NAOJ/NRAO)/PHANGS, S. Dagnello (NRAO).

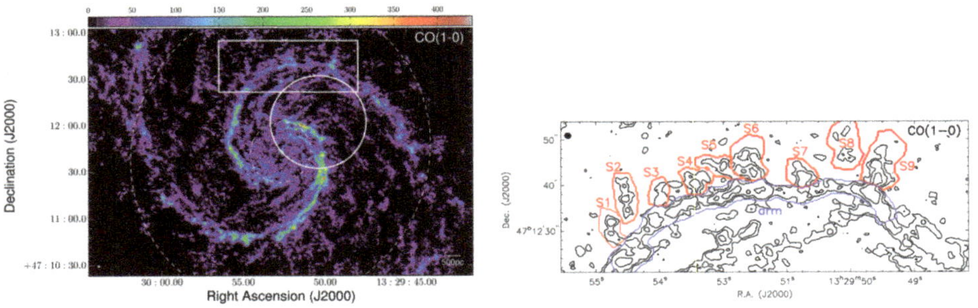

Figure 9.11. The Plateau de Bure interferometer (PdBI) for the PdBI Arcsecond Whirlpool Survey (PAWS; Schinnerer et al. 2013). On the left is the central 9 kpc of M51 in exquisite detail. Not only can you see the molecular clouds that follow the spiral arm structure, hosting stellar nurseries but also the lower density details in the disk itself. On the right is a detail from the image on the right (white box) with a spiral arm and the "feathering" structure that the ISM in M51 exhibits. Figure from Schinnerer et al. (2017).

Hughes et al. 2013; Schinnerer et al. 2013, 2017; Colombo et al. 2014a, 2014b; Meidt et al. 2015), Figure 9.11.

The carbon monoxide view of nearby galaxies offers a way to trace the molecular gas component, together with the atomic hydrogen reservoir and the dust that shields the molecular phase, one can take a detailed inventory of the cold gas phases in the disks of galaxies and compare these directly to the star formation that occurs. Surveys such as PAWS and the PHANGS-ALMA (below) promise to reveal the different physical conditions in which star formation occurs in different types of disk galaxies.

9.3.2 ALMA

The Atacama Large Millimeter Array is the next game-changing instrument in the sub-mm wavelength regime. Rather than a single (space) telescope such as Herschel, it is a interferometer, similar to the JVLA or MeerKAT. As a result, it can resolve these wavelengths in much greater detail. The real challenge is to encompass the entire galaxy when studying nearby galaxies. Figure 9.10 shows a compilation of nearby galaxies from the PHANGS—Physics at High Angular Resolution in Nearby GalaxieS project (Lee et al. 2021), a project to map nearby galaxies with Hubble. PHANGS-ALMA (Sun et al. 2018, 2020; Schinnerer et al. 2019; Chevance et al. 2020; Lang et al. 2020; Anand et al. 2020; Rosolowsky et al. 2021; Lee et al. 2021; Williams et al. 2021; Leroy et al. 2021) accompanies this large HST project with dust and gas maps in incredible detail. Combined in Figure 9.12; it shows both the newly formed and the dusty gas clouds from which stellar nurseries spring.

This makes the study of the physical processes of star formation in a range a different galaxies possible. As a mild caution, one should note that interferometers have a range of scales that they are sensitive for: the high resolution that resolved the clouds that accompany stellar nurseries means that galaxy-scale diffuse emission is removed in the images in Figure 9.10.

Figure 9.12. A color composite of the Hubble ultraviolet and optical observations (white) and the ALMA sub-mm observations (orange). Credit: ALMA (ESO/NAOJ/NRAO)/PHANGS, S. Dagnello (NRAO).

An embarrassment of riches: You may notice that certain galaxies get all the attention of telescopes. Especially smaller surveys (e.g., less than a few dozen galaxies) end up looking at the same galaxies over and over again. This is a pretty well-known phenomena where certain objects got good data with one telescope early and thus became the natural follow-up target for other telescopes and wavelengths. Good data already exists, so no need to get new. This does come with a built-in weakness: one is now completely reliant on the original survey design which hopefully is representative enough. Good examples are the SINGS (Kennicutt et al. 2003) and THINGS (Walter et al. 2008), and KINGFISH (Kennicutt et al. 2011) sample. It is now a classical reference set of galaxies but who can tell me what the selection criteria were?

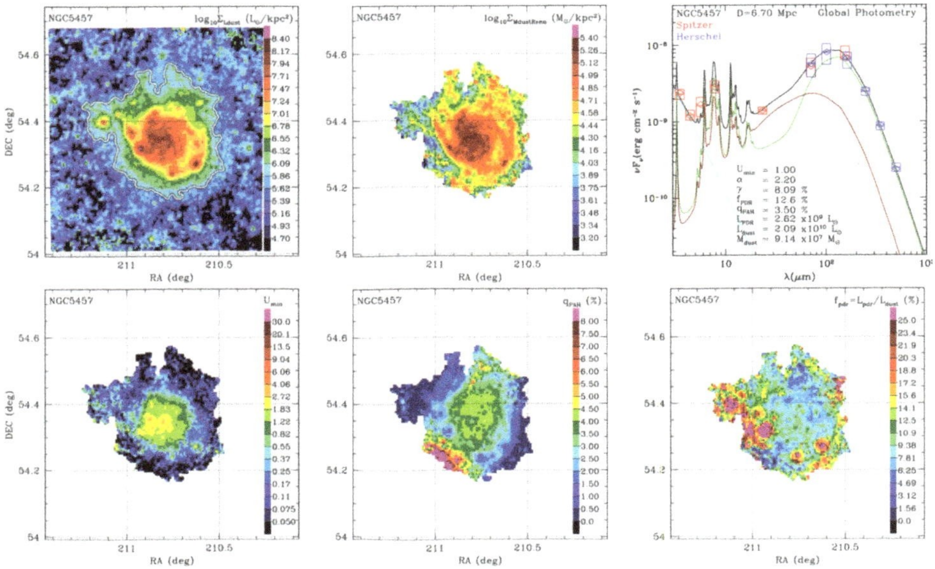

Figure 9.13. NGC5457 (M101) dust model results at the 160 μm resolution of Herschel. Dust luminosity per area Σ_{Ld} (top left) is shown for entire field, with adopted galaxy mask boundary in white. Dust mass per area Σ_{Md} (top middle) is after renormalization. Stellar irradiation (U_{min} , bottom left), fraction of PAH molecules (q_{PAH}, bottom middle) and fraction Photodissociation region (f_{PDR}, bottom right). The global SED (top right) is shown for single-pixel modeling, with contributions from dust heated by U_{min} (green), dust heated by $U > U_{min}$ (red) and starlight (cyan); values of U_{min} and M_d in the figure label are for the Draine & Li model (Draine & Li 2007) before renormalization. Herschel (blue rectangles) and Spitzer (red rectangles) photometry is shown. Diamonds show the band-convolved flux for the model. Figure reproduced from Aniano et al. (2020).

Figure 9.13 illustrates what all is involved to interpret dust morphology. It shows a single filter (160 μm on Herschel) but to interpret it, one needs to model the incident light on the dust at each spot (U), the fraction of PAHs (f_{PAH}) before one can estimate the surface density of dust in the galaxy (in this case M101 again). As one can image, this is quite a bit more involved than estimating gas surface density from 21 cm emission line (a direct tracer) or stellar density from a near-infrared filter image (a single mass-to-light ratio). It also marks the shift how one interprets galaxy imaging data. Rather than describing the morphology, one models the local conditions: stellar population and its radiation field, the composition and temperature of the interstellar medium etc.

Still, one can apply more traditional methods of morphological analysis on single filter images of the dust in galaxies (see below) and this too can perhaps reveal the underlying processes that shape the appearance of the galaxy in all the different wavelengths. We will skip optical and near-infrared at this point since the rest of books is effectively about this.

9.4 Ultraviolet

Ultraviolet emission from galaxies was first explored in detail for statistical numbers of galaxies using GALEX satellite. We already covered eXtended Ultra-Violet disks in Chapter 7. Ultraviolet and Hα emission is linked to recent massive star formation because UV is where young O- and B-star spectra peak and ionizing hydrogen occurs in close proximity to ultraviolet sources. So often the ultraviolet and Hα morphologies of galaxies will resemble each other fairly closely. The timescales involved for recent star formation depends a little on circumstances, with commonly quoted timescales being 100 Myr for FUV and 10 Myr for Hα, although these scales may well be shorter (Flores Velázquez et al. 2021).

Ultraviolet is patchier and more extended in spiral galaxies, making is much more difficult to identify which discrete source is part of a galaxy and which are background sources of some kind.

9.4.1 UV "Frosting"

Ultraviolet light does not just originate from spirals alone. Many elliptical galaxies show a little UV light, either from a "frosting" of star formation (they are not *completely* dead) or the "UV-upturn," ultraviolet emission that cannot be explained by their stellar population unless one includes flaring from older stars, Figure 9.14.

9.4.2 UV Vertical Profiles

The vertical profile of galaxies show another interesting effect in the ultraviolet: scatter off the cirrus dust clouds high above and below the plane of the disks of galaxies. Figure shows NGC 891 observed with GALEX (Seon et al. 2014), where

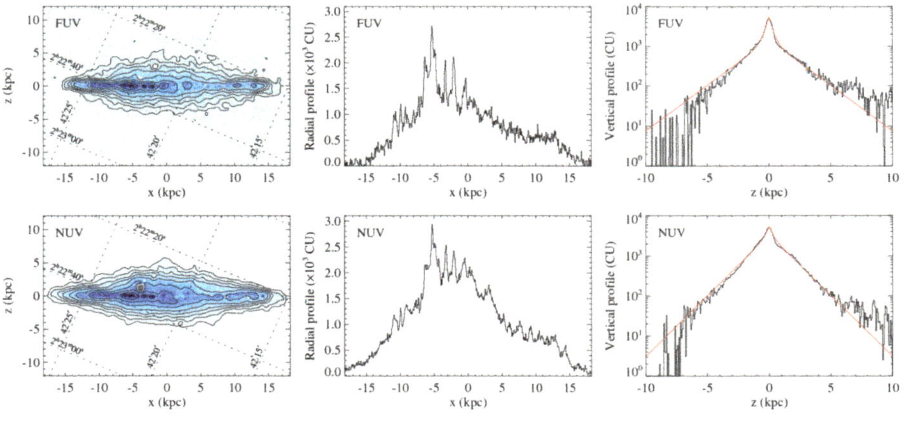

Figure 9.14. GALEX maps and vertical profiles of NGC 891. Concentric contours and white regions are artifacts due to the masking of foreground stars. A distance of 9.5 Mpc to NGC 891 was assumed. Black solid lines represent the observed profiles. The red curves are the best-fit two-exponential functions for the vertical profiles. The vertical profiles were obtained by averaging over the whole disk and the radial profiles over $|z| < 5$ kpc. Figure reproduced from Seon et al. (2014).

additional light above and below the plane is not due to *in situ*[4] star formation but light from the disk scattered into our line-of-sight by dust clouds high above and below this disk.

A part of the ultraviolet component will always be scatter off cirrus clouds above and below the disk of galaxies. In a few cases, GALEX or Hα observations already show evidence for scatter of high latitude dust (Seon & Witt 2012; Seon et al. 2014; Seon & Draine 2016; Seon 2018; Shinn & Seon 2015; Shinn 2019).

9.5 X-Ray

Galaxies can and are observed in X-ray, revealing the most energetic components of the galaxy's ecosystem (binaries, black holes etc). Observatories such as the Chandra and XMM-Newton space observatories have steadily returned data. However, morphology is very tricky as (a) count rates are often low, (b) the energetic sources in a galaxy are often discrete and (c) the spatial resolution of X-ray observatories is not really comparable with other wavelengths. Figure 9.15 shows NGC 891 observed with Chandra (Tyler et al. 2004; Strickland et al. 2004).

However, we know that a fraction of the gas that falls into the gravitational well that is a galaxy is heated to tens of thousands of degrees Kelvin, emitting X-ray light. Some of the gas ejected from the disk is equally ionized and hot. So there should be diffuse halo of X-rays around most galaxies. It is just difficult to detect. Thus, X-ray

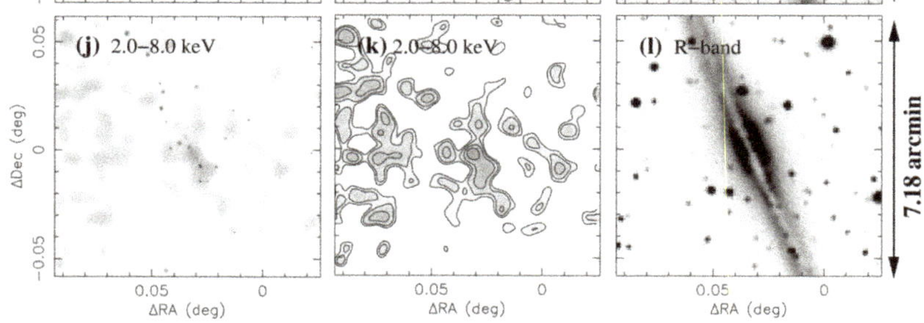

Figure 9.15. NGC 891 observed in X-ray, expressed as energies. Hα and R-band images are to show where one would expect star formation driven sources and what the stellar component of this galaxy look like. Figure reproduced from Strickland et al. (2004). © 2004. The American Astronomical Society. All rights reserved.

[4] Latin for "in place" because we astronomers are fancy like that.

galaxy morphology is a collection of energetic (unresolved) point sources with a diffuse underlying component.

9.6 Sérsic Profile

In the optical and near-infrared, one can characterize the profiles of galaxies using Sérsic functions (Chapter 2) and explore the dependence on wavelength where the profile is fit. Different stellar populations contribute different fractions of the light at a given wavelength (e.g., O- and B-stars dominate blue, the bulk of low-mass stars dominate red wavelengths). Once outside the wavelengths dominated by stellar

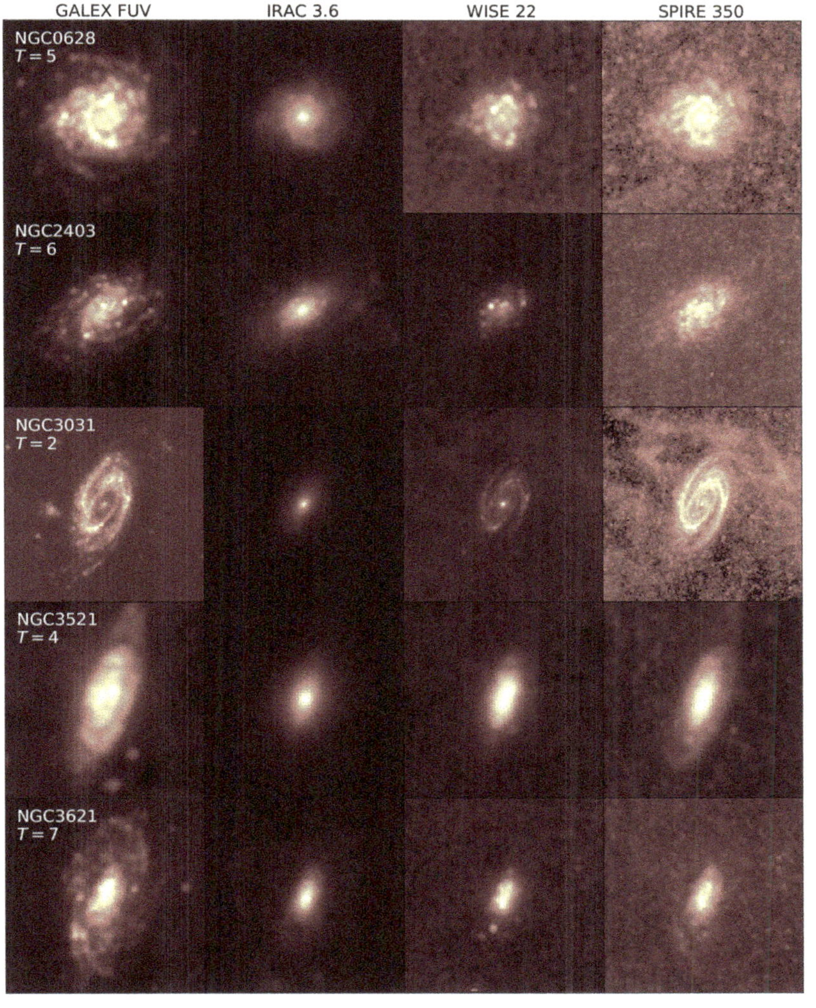

Figure 9.16. GALEX FUV, IRAC 3.6 μm, WISE 22 μm and SPIRE 350 μm images for NGC0628 (M 74), NGC2403, NGC3031 (M 81), NGC3521, and NGC3621. All images are convolved to the same resolution and shown on the same brightness scale. Figure reprinted with permission from Baes et al. (2020) © ESO.

emission however, the Sérsic profile is not as good a descriptor of the general appearance. For example, HI disks or UV disks are not well described by these.

9.7 Morphometrics

Similar to the Sérsic profile, the galaxy morphometrics change as they are evaluated from ultraviolet to the sub-mm (Baes et al. 2020). Figure 9.16 shows a few examples of nearby galaxies for which high-resolution data is available across wavelength. One can then calculate the morphometrics (see Chapter 6 for each image at each wavelength as well as one for the stellar mass, star formation and dust maps that can be produced from spectral energy distribution fitting). Figure 9.17 shows the values of half-light radius, asymmetry, Gini, concentration, smoothness and M_{20} for every image taken at wavelengths of 0.1–300 micron, and the stellar mass, star formation, and dust maps that were produced by the models of these galaxies. There is a fair amount of change with wavelength but plots such as these can motivate how to compare different wavelength morphologies: which wavelength image's morphometrics come the closest to the physical parameter map one is interested in? In other words, which image should be used as the proxy for stellar mass morphology? or star formation?

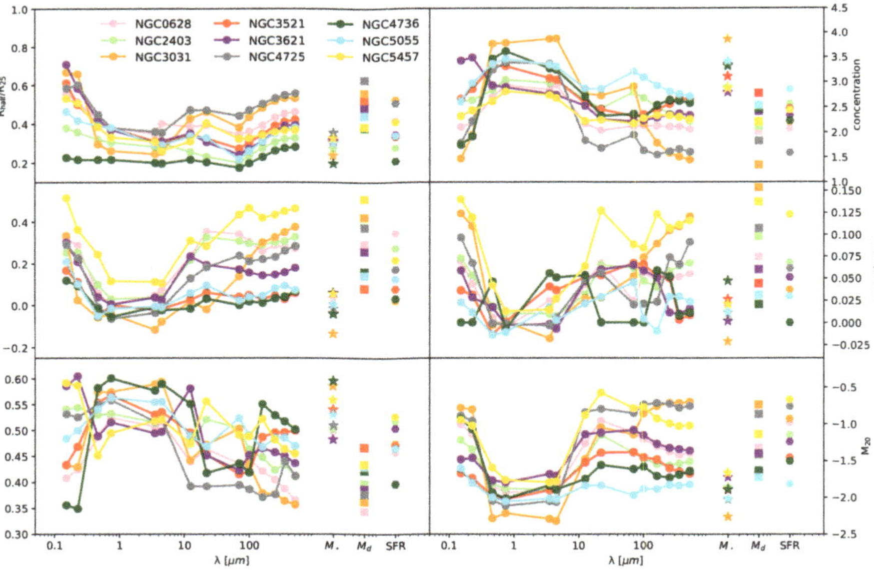

Figure 9.17. Morphometric parameters for the galaxies in Figure 9.16 and several more as a function of wavelength, and for the stellar mass (M), dust mass (M_d), and star formation rate (SFR) maps. The half-light radius (R_{half} or R_e) normalized by the optical R_{25} diameter (top left, see chapter 3), the concentration index (top right), the asymmetry index (middle left), the smoothness index (middle right), the Gini coefficient (bottom left), and the M_{20} indicator (bottom right). Wavelengths by dominated by star formation are in the UV and also 22 μm. Stellar mass in the near-infrared (3.6 and 4.6 μm) and dust in the far-infrared and sub-mm. But note their morphometrics do not fully correspond to the map of the physical process they are dominated by. Figure reprinted with permission from Baes et al. (2020) © ESO.

9.8 Assignments

The dependence of morphology on wavelength features in "Virgo cluster and the morphology–density relation" (jupyter notebook 15) but it also touches topics that are discussed in Chapter 10.

References

Anand, G. S., Lee, J. C., Van Dyk, S. D., et al. 2020, MNRAS, 501, 3621
Aniano, G., Draine, B. T., Hunt, L. K., et al. 2020, ApJ, 889, 150
Baes, M., Nersesian, A., Casasola, V., et al. 2020, A&A, 641, A119
Boomsma, R., Oosterloo, T. A., Fraternali, F., et al. 2005, A&A, 431, 65
Chevance, M., Kruijssen, J. M. D., Hygate, A. P. S., et al. 2020, MNRAS, 493, 2872
Colombo, D., Meidt, S. E., Schinnerer, E., et al. 2014a, ApJ, 784, 4
Colombo, D., Hughes, A., Schinnerer, E., et al. 2014b, ApJ, 784, 3
de Blok, W. J. G., Adams, E. A. K., Amram, P., et al. 2016, in Proc. MeerKAT Science: On the Pathway to the SKA, 7
de Blok, W. J. G., Józsa, G. I. G., Patterson, M., et al. 2014, A&A, 566, A80
Draine, B. T., & Li, A. 2007, ApJ, 657, 810
Flores Velázquez, J. A., Gurvich, A. B., Faucher-Giguère, C.-A., et al. 2021, MNRAS, 501, 4812
Gentile, G., Józsa, G. I. G., Serra, P., et al. 2013, A&A, 554, A125
Giese, N., van der Hulst, T., Serra, P., & Oosterloo, T. 2016, MNRAS, 461, 1656
Heald, G., Józsa, G., Serra, P., et al. 2011a, A&A, 526, A118
Heald, G., Allan, J., Zschaechner, L., et al. 2011b, in IAU Symp. 277, ed. C. Carignan, F. Combes, & K. C. Freeman, 59
Heald, G. 2015, in IAU Symp. 309, Galaxies in 3D across the Universe, 69
Hibbard, J. E., van Gorkom, J. H., et al. 2001, in ASP Conf. Ser. 240, Gas and Galaxy Evolution, 657
Holwerda, B. W., Pirzkal, N., de Blok, W. J. G., et al. 2011a, MNRAS, 416, 2401
Holwerda, B. W., Pirzkal, N., Cox, T. J., et al. 2011b, MNRAS, 416, 2426
Holwerda, B. W., Pirzkal, N., de Blok, W. J. G., et al. 2011c, MNRAS, 416, 2437
Holwerda, B. W., Pirzkal, N., de Blok, W. J. G., et al. 2011d, MNRAS, 416, 2401
Holwerda, B. W., Pirzkal, N., de Blok, W. J. G., et al. 2011e, MNRAS, 416, 2415
Holwerda, B. W., Pirzkal, N., de Blok, W. J. G., et al. 2011f, MNRAS, 416, 2447
Hughes, A., Meidt, S. E., Schinnerer, E., et al. 2013, ApJ, 779, 44
Hunter, D. A., Elmegreen, B. G., & Gehret, E. 2016, AJ, 151, 136
Hunter, D. A., Ficut-Vicas, D., Ashley, T., et al. 2012, AJ, 144, 134
Ianjamasimanana, R., Walter, F., de Blok, W. J. G., Heald, G. H., Brinks, E., et al. 2018, AJ, 155, 233
Kamphuis, J., Sancisi, R., & van der Hulst, T. 1991, A&A, 244, L29
Kennicutt, R. C., Armus, L., Bendo, G., et al. 2003, PASP, 115, 928
Kennicutt, R. C., Calzetti, D., Aniano, G., et al. 2011, PASP, 123, 1347
Lang, P., Meidt, S. E., Rosolowsky, E., et al. 2020, ApJ, 897, 122
Lee, J. C., Whitmore, B. C., Thilker, D. A., et al. 2021, arXiv:2101.02855
Leroy, A. K., Schinnerer, E., Hughes, A., et al. 2021, arXiv:2104.07739
Marasco, A., Fraternali, F., Heald, G., et al. 2019, A&A, 631, A50
Meidt, S. E., Hughes, A., Dobbs, C. L., et al. 2015, ApJ, 806, 72

Meidt, S. E., Schinnerer, E., García-Burillo, S., et al. 2013, ApJ, 779, 45
Pingel, N. M., Pisano, D. J., Heald, G., et al. 2018, ApJ, 885, 36
Ramatsoku, M., Murgia, M., Vacca, V., et al. 2020, A&A, 636, L1
Reynolds, T. N., Westmeier, T., Staveley-Smith, L., Chauhan, G., & Lagos, C. D. P. 2020, MNRAS, 493, 5089
Richter, P., Winkel, B., Wakker, B. P., et al. 2018, ApJ, 868, 112
Rosolowsky, E., Hughes, A., Leroy, A. K., et al. 2021, MNRAS, 502, 1218
Schinnerer, E., Hughes, A., Leroy, A., et al. 2019, ApJ, 887, 49
Schinnerer, E., Meidt, S. E., Colombo, D., et al. 2017, ApJ, 836, 62
Schinnerer, E., Meidt, S. E., Pety, J., et al. 2013, ApJ, 779, 42
Seon, K. I. 2018, ApJ, 862, 87
Seon, K. I., & Draine, B. T. 2016, ApJ, 833, 201
Seon, K. I., & Witt, A. N. 2012, ApJ, 758, 109
Seon, K. i., Witt, A. N., Shinn, J.-h., & Kim, I.-j. 2014, ApJL, 785, L18
Shinn, J. H. 2019, MNRAS, 489, 4690
Shinn, J. H., & Seon, K. I. 2015, ApJ, 815, 133
Strickland, D. K., Heckman, T. M., Colbert, E. J. M., Hoopes, C. G., & Weaver, K. A. 2004, ApJS, 151, 193
Sun, J., Leroy, A. K., Ostriker, E. C., et al. 2020, ApJ, 892, 148
Sun, J., Leroy, A. K., Schruba, A., et al. 2018, ApJ, 860, 172
Tyler, K., Quillen, A. C., LaPage, A., & Rieke, G. H. 2004, ApJ, 610, 213
Walter, F., Brinks, E., de Blok, W. J. G., et al. 2008, AJ, 136, 2563
Williams, T. G., Schinnerer, E., Emsellem, E., et al. 2021, AJ, 161, 185
Zschaechner, L., et al. 2011a, AAS Meeting 217, 246.23
Zschaechner, L. K., Rand, R. J., Heald, G. H., Gentile, G., & Józsa, G. 2012, ApJ, 760, 37
Zschaechner, L. K., Rand, R. J., Heald, G. H., Gentile, G., & Kamphuis, P. 2011b, ApJ, 740, 35

Chapter 10

Morphology Evolution

Galaxies evolve individually and as a population. Individual galaxies can be followed over the billions of years they take to evolve by either simulating them and speeding up time or one can identify the remnants of significant events in its history from remnants of the stellar populations (galaxy archeology). To observe the general evolution of galaxies, one needs to identify relations in the local universe and then examine how these relationships look in the distant universe where light travel time gives us a look back into the past. Were there elliptical galaxies at every epoch? How does the balance of morphologies change over time? Does the Hubble tuning fork work at every time during the age of the universe?

Starting from the Big Bang and a universe expanding and filled with hydrogen and dark matter, we expect stars to form at some point in the past. In the local universe, we see a myriad of galaxies in a riot of colors. Studying the evolution of galaxies aims to see how the universe got from one state to the latter.

10.1 Observed Dependencies

There are two trends with broad morphology that have stood out early: disk galaxies are bluer, and spheroidals are red. Second, the more crowded in galaxies are, the more spheroidals are present. These are the morphology–color and morphology–density relations.

The morphology–color relation has been noticed early from photographic plates and came to splendid light in digital images of galaxies. So much so, we expect disk galaxies to always be bluer than spheroidal center or spheroidal galaxies. Yet, there are red disk galaxies (Masters et al. 2010; Rowlands et al. 2012) and blue spheroidals. This is a 90% rule that disks are blue. Not all of them need to be. Using kinematics in addition to morphology measures and color, it becomes clear there are rotating very disk-like "spheroidals."

Similarly, it was noted early that in the dense environments of galaxy clusters, the fraction of disk and irregular galaxies near the center of the cluster was much lower,

doi:10.1088/2514-3433/ac2c7dch10

while the fraction of ellipticals went up steeply (Houghton 2015). This change in morphology (and because of the above relation between morphology and broad color) with color and the density of the galaxy's environment is colloquially known as the Butcher–Oemler effect. Figure 10.1 recreated the original illustration showing higher fractions of spheroidals with local density. However, as soon as one moves out of a cluster and its environs, this density-morhology (or color) relation is a lot harder to show. The effect in groups, rather than clusters is much harder to quantify. Does the rate of star formation goes up (bluer galaxies) in groups or not? This is proving trickier (Prescott et al. 2011; Wijesinghe et al. 2012; Grootes et al. 2012; Christodoulou et al. 2012). One thing does seem to be plausible: the morphological transformation does not happen exclusively in the cluster environment. Clusters form from groups of galaxies falling in, and in those groups "pre-processing" appears to happen; these galaxies were transitioning to spheroidals and lower star formation rates before they entered the cluster.

10.2 Morphology Evolution

In broad terms, we see fewer spheroidal galaxies at higher redshifts. The elliptical galaxies are an end-stage (well the current end-stage) of galaxy evolution. Star

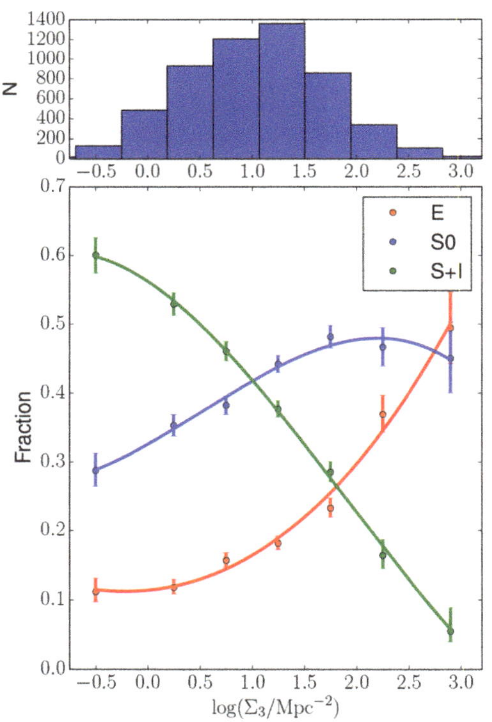

Figure 10.1. Reproduction of the morphology–density relation of D80 using up-to-date cluster redshifts, projected densities to the third nearest neighbor and field-edge or completeness corrections. Figure reproduced from Houghton (2015).

formation in a given volume of galaxies peaked at $z \sim 1.5$ (7 Gyr ago; Madau & Dickinson 2014) and all galaxies have relatively been forming stars at a lower rate since. Figure 10.2 shows the evolution of elliptical/spheroidal fraction and disk galaxies with redshift for massive galaxies. For those we know we are complete, the much less massive galaxies (and therefore fainter) are harder to tally. Fraction of disk galaxies goes down, spheroidals goes up. At some point in the distant past however, the argument that we are looking at either a disk or a spheroid becomes a bit moot because the entire galaxy observed is (a) barely resolved, (b) the size of a big

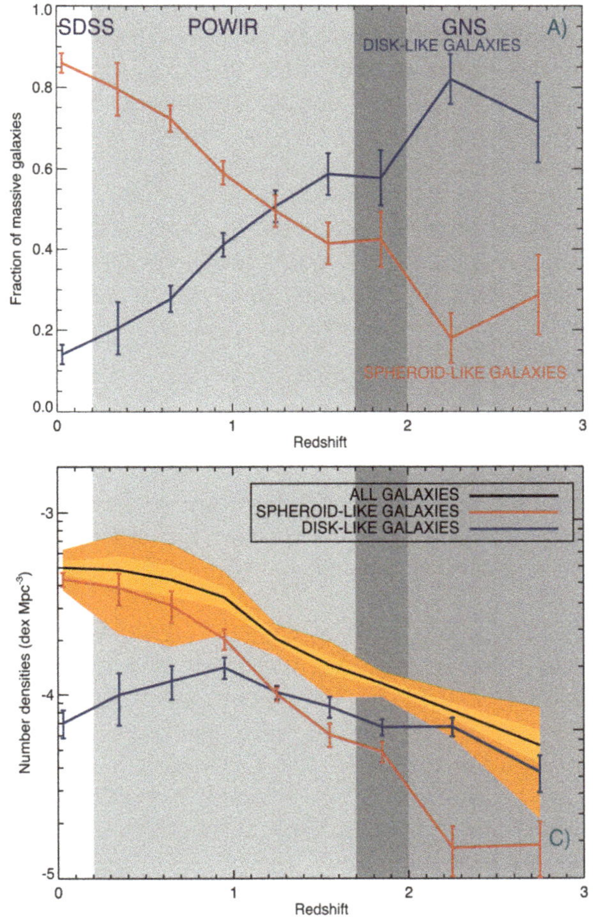

Figure 10.2. Fraction of massive ($M_* > 10^{11}M_\odot$) galaxies showing disk-like surface brightness profiles (Sérsic index, $n < 2.5$) and spheroid-like ones ($n > 2.5$) as a function of redshift. Different color backgrounds indicate the redshift range expanded for each survey: SDSS, POWIR/DEEP2, and GNS. Top panel shows the absolute numbers of disks (blue) and spheroids (red), bottom panel shows the density, corrected for the smaller volumes in the universe. The total density of galaxies is shown as the orange shaded area. Figure adapted from Buitrago et al. (2013). © 2012 The Authors, Published by Oxford University Press on behalf of the Royal Astronomical Society.

HII region (see Figure 3.10), and (c) a whorl of star formation. But the disk/spheroid morphology establishes itself fairly early ($z = 3$, 10+ Gyr ago).

10.2.1 K-correction of Morphology

How to observe morphology of galaxies over the billions of years that the universe has existed? There are two ways: galaxy archeology, where one identifies the populations and generations of stars that make up contemporary, nearby galaxies. The GHOSTS survey work is part of that approach, as is the detailed examination of Andromeda. The other way is to use telescopes as time machines: observe much more distant galaxies and compare them to nearby ones. The light from the more distant galaxies has traveled for billions of years and therefore the galaxies have the appearance they had billions of years ago. The distant galaxies are redshifted with the expansion of the universe. The light they emitted has shifted to longer wavelengths. Similarly, the galaxy surface brightness has dimmed by distance proportionally with the redshift $(1 + z)^2$, and because it is more distant, our best telescopes can only discern the larger morphological details.

To emulate that, one can take the bluer filter observed locally, dim appropriately and smooth the resulting image to match the best we can do at high redshift. This is call "morphological K-correction." K-correction is the term used where one adds or removes flux to the luminosity or color observed at one redshift to be able to compare to a galaxy at another redshift. (This is explained nicely in Chris Conscelice's book).

10.2.2 Examples: Bars and Dust Lanes

Suppose there is an obvious morphological feature that you want to see if the population of galaxies at different epochs has more or less of over the course of the evolution of galaxies over time. We will examine two examples here: the fraction of galaxies that has a clear central bar and the fraction of edge-on galaxies that have a dust-lane.

First it is useful to estimate out to how far the highest resolution images one can obtain could identify the morphological feature of interest. The Hubble Space Telescope remains the gold standard, although ground-based telescopes and ALMA can do better on occasion. The Hubble has a resolution of about 0.1 arcseconds. This translates to 1 kpc out to $z = 0.8$ (6.6 Gyrs ago).

For the past 7 billion years, we can reasonably expect to identify morphological features greater than 1 kpc. So nuclear star clusters or central bars are out but big bars or a clear dust lane are doable.

10.2.2.1 Evolution of Bars

What one now does is take optical SDSS images and classify those and then morphologically K-correct them to what Hubble would see at redshifts out to $z = 0.8$. The Hubble images are the most expensive so you only have a single filter (Figure 10.3). This is the case for the COSMOS survey (Scoville et al. 2007) of a single field. This is the biggest single field observed with Hubble to characterize the

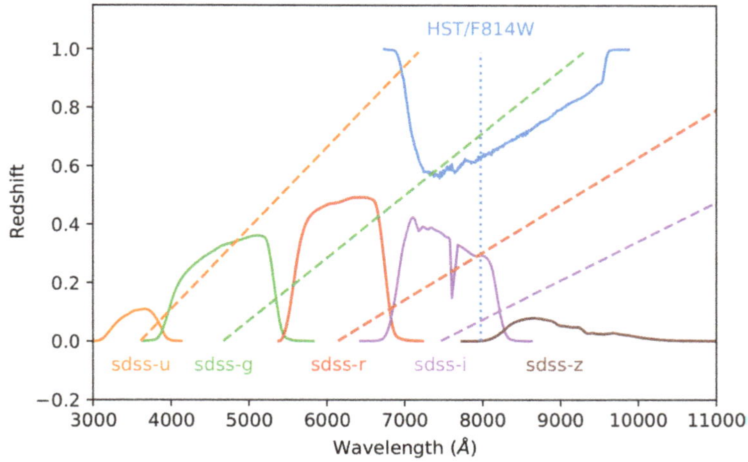

Figure 10.3. The five Sloan Digital Sky Survey filters (sdss-*ugriz*) compared to the Hubble Space Telescope's F814W filter (top, inverted). The central effective wavelengths of each filter is shown as dashed lines for the Sloan filters and a dotted line for the HST. The Sloan filters are redshifted, intersecting where HST is observing at different wavelengths; this is the *rest-frame wavelength* of the HST observation at that redshift.

morphology of galaxies at intermediate redshifts. Figure 10.4 shows an example of barred galaxies in SDSS that have been redshifted to higher redshift using the morphological K-correction. They have been dimmed, smoothed, and the sdss filter was chosed to correspond to the *restframe wavelength* of the light emitted that now has redshifted to the HST filter.

10.2.2.2 Dust Lanes

Following a similar approach as the bar fraction search, one can look for dust lanes in edge-on galaxies. This is a little more ambitious as a dust lane in the middle of an edge-on disk is only about a kpc thick. Figure 10.5 shows how the fraction of dust lanes in massive edge-on galaxies remains at ~80% no matter what redshift. A lack of evolution such as Figure 10.5 do constrain how we understand galaxies, for example as a balance between surface gravity and turbulent input into the ISM (Dalcanton et al. 2004).

10.3 Morphology Evolution

We see galaxy populations evolve morphologically. The general trend (after corrections for volume and incompleteness) is that spiral and disk galaxies are on the decline and relatively more elliptical or spheroidal galaxies are being made. This is the first evidence that any morphological evolution is from disks to spheroids. Figure 10.2 shows the relative numbers of disks versus spheroid galaxies (according to their Sérsic fits, see Chapter 2) and their relative densities in a cosmic volume as a function of redshift (lookback time). At higher redshift, earlier in the cosmic evolution of galaxies, there were in a given volume of space, more disk-like galaxies

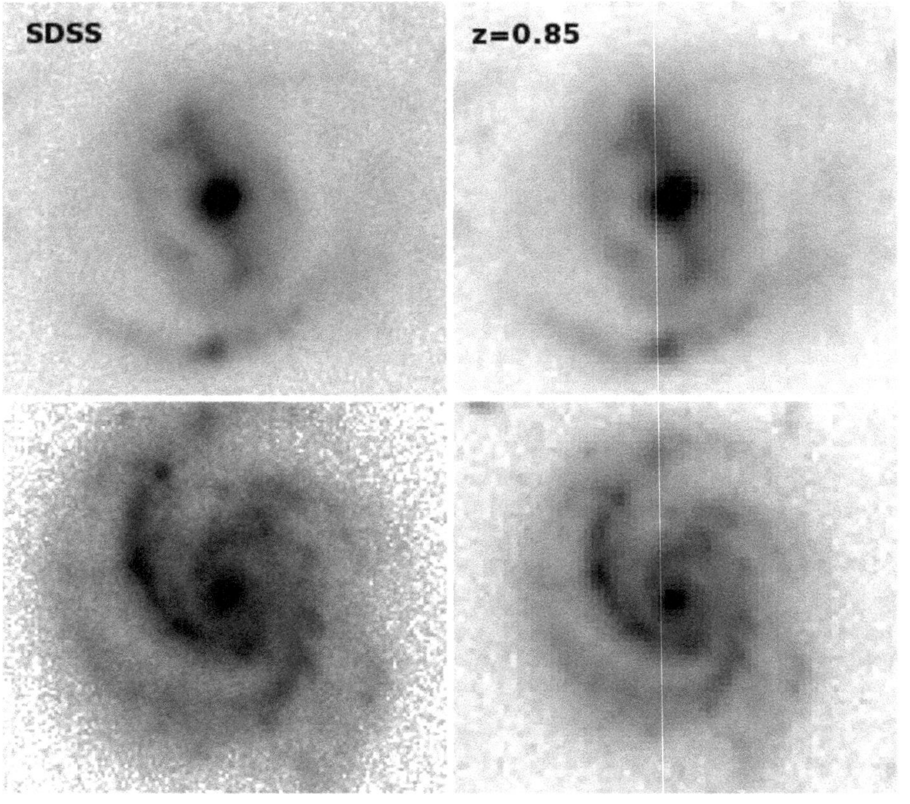

Figure 10.4. Two examples of artificially redshifted SDSS galaxies (Sheth et al. 2008). A barred spiral is shown in the top row and an unbarred spiral galaxy in the bottom row. The left column shows the original SDSS g-band image and the right column is the artificially redshifted image at $z = 0.84$. Figure reproduced from Sheth et al. (2008).

at $z \sim 3$ (some 10 Billion years ago) then there were in the present day ($z \sim 0$ is the here and now, give or take a billion years).

10.4 Mergers

One dramatic change in morphology is when two galaxies coalesce after a tidal dance. Famously, the Andromeda galaxy and our own Milky Way will have such an encounter. Over the course of billions of years, the two galaxy will orbit and eventually merge, dramatically changing their appearance as stellar orbits are randomized and gas clouds collide to trigger star formation. Partly because the morphologial effects are so dramatic, it was long thought that this was the only way to produce big elliptial galaxies. The fact that we see more ellipticals in dense environments also points to the role mergers (may) play in morphological transformation.

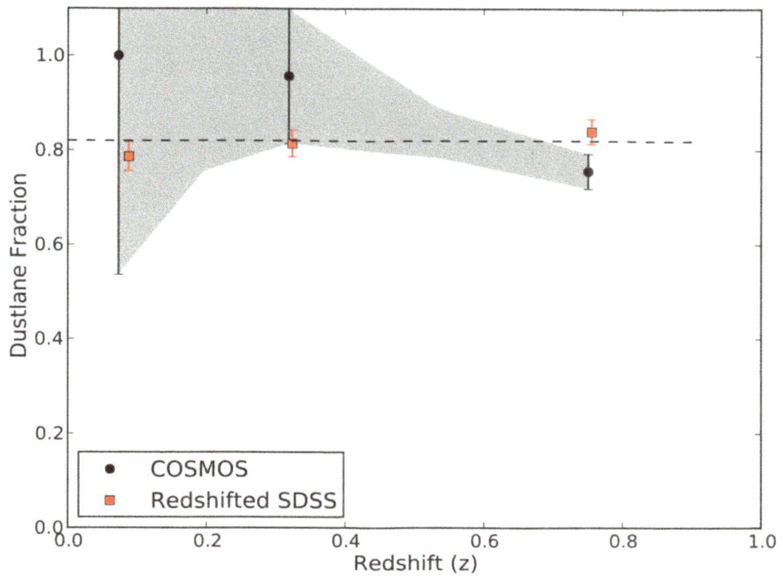

Figure 10.5. The fraction of edge-on galaxies with a visible dust-lane. The fraction (and uncertainty) in the original COSMOS galaxies. To compare, the SDSS-g, r, and i edge-ons are redshifted and the fraction of dust lanes that can be identified. The observed and artificially redshifted populations are both consistent with 80% of edge-on galaxies having a visible dust lane since $z \sim 0.8$ or 6 billion years go. Figure reproduced from Holwerda et al. (2012). © 2012. The American Astronomical Society. All rights reserved.

Without a doubt, when two galaxies are merging, the resulting morphology is dramatic. Figure 10.6 shows a galaxy galaxy observed with the Hubble Space Telescope.

The exact importance of galaxy mergers is being debated right now. We typically speak of major mergers when both galaxies are close in mass (1:1 mass ratio down to 1:3 mass ratio). Minor mergers are when there is a clear massive and satellite galaxy, for mass ratios greater than 1:3.

10.4.1 Spirals to Ellipticals

Simulations of two spirals merging dramatically, as visible in Figure 10.6, include not just a redistribution of the angular momentum of the stellar orbits but also a lot of new star formation during both coalescence and close encounters due to tidal forces. This is the kind of merger, one hopes to trace across epochs using morphology. Not only is the morphology affected by the tidal forces, there is also additional stellar flux in the UV and blue light thanks to fresh star formation. Searches like those illustrated in Figure 10.18 are an example of characterizing how frequent mergers like these occur and therefore how important they are in galaxy evolution. Because there is a gas collision and subsequent starburst component to these kinds of mergers, they are often referred to as "wet" mergers.

Figure 10.6. A gallery of tidally interacting galaxies observed with the Hubble Space Telescope. The morphological transformation of each of these galaxies is dramatic. Credit: NASA/ ESA/ STScI/AURA (The Hubble Heritage Team)—ESA/Hubble Collaboration/ University of Virginia, Charlottesville, NRAO, Stony Brook University (A. Evans)/ STScI (K. Noll)/ Caltech (J. Westphal), from ESA: http://www.esa.int/Science_Exploration/Space_Science/Plethora_of_interacting_galaxies_on_Hubble_s_birthday.

10.4.2 Big Ellipticals

But the two galaxies that merge need not be two disk galaxies. If two elliptical galaxies merger, the effect may not include a burst of star formation, nor such clear deformation of the morphology but it must happen, especially if one considers that ellipticals are more common in dense environments (clusters of galaxies). The morphological drama as a result if often at lower surface brightness (e.g., shells and tidal tails) but no less clear or dramatic. The brightest cluster galaxy (BCG), the monster galaxy often found in the kinematic center of clusters often shows evidence of multiple nuclei, meaning that it is really a clump of half-merged ellipticals, still in the process of fully merging into a new elliptical. Because of the lack of starbursts, these elliptical–elliptical galaxy mergers are referred to as "dry" mergers.[1]

10.5 Galaxies in the Universe

How many galaxies are ellipticals and how many are spirals at a given time during the evolution of galaxies? The best way to answer that is to compare volumes of

[1] This colloquial way of referring to mergers implies that mergers in between these two scenarios are "moist" but that word is just wrong if it doesn't describe chocolate cake.

galaxies nearby and further away (and therefore earlier in their evolution) and count how many of each morphology one encounters. Is the brightest of each kind at the same luminosity? How many of the smaller (less massive) galaxies are there? Key is to account (and possibly correct) for those galaxies that are missing out of your tally because of depth (not observed because galaxy too faint) or volume (rare, brightest galaxy not included because the volume of the survey was too small).

10.5.1 Stellar Mass in Morphological Type

In the a representative volume of the nearby universe, one can count the relative numbers of galaxies, identifying how many disks and spheroids there are at a given stellar mass. One can take that even a step further and decompose all the disk galaxies into disks and bulges. That way, the total stellar mass in spheroidal structures (pressure supported, see Chapter 8) versus all the stars in disks (or rotationally supported, see Chapter 8). Figure 10.7 shows the contributions of different Hubble Types in GAMA (Liske et al. 2015) to the numbers in a cosmological volume as a function of galaxy stellar mass (M_*). Number densities as a function of luminosity of mass are often described using a Schechter function:

For luminosities this can be expressed in magnitudes:

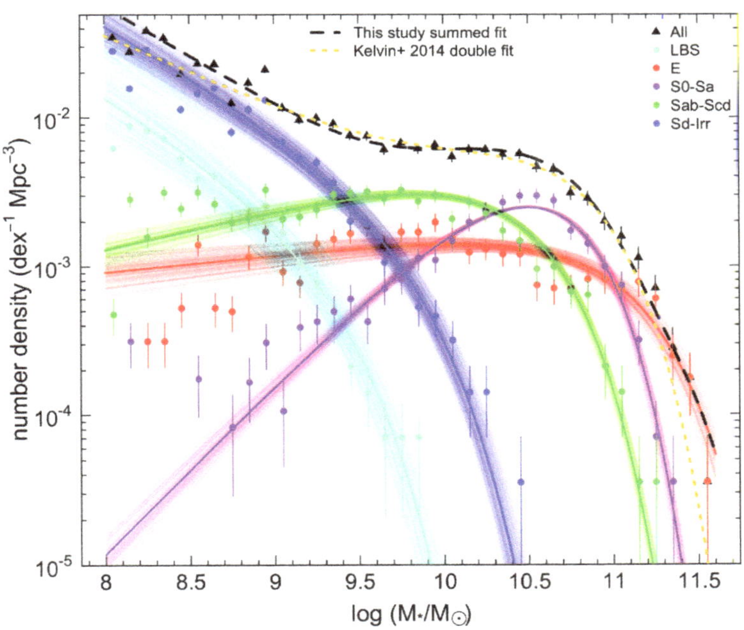

Figure 10.7. The stellar mass function, i.e., the number density in a cosmological volume as a function of stellar mass, broken down by galaxy morphological type. The points are data from the Galaxy And Mass Assembly (GAMA; Liske et al. 2015) survey. Lines are fits to the data using a single Schechter function. The total of all these is combined (black dashed line) and compared against the total found (Kelvin et al. 2014). Figure reproduced from Moffett et al. (2016a). © 2016 The Authors, Published by Oxford University Press on behalf of the Royal Astronomical Society.

$$\phi(M) = 0.4 \ln 10\phi^{*}[10^{0.4(M^{*}-M)}]^{1+\alpha} \exp[-10^{0.4(M^{*}-M)}] \qquad (10.1)$$

or in luminosities:

$$\Phi(L) = \left(\frac{\phi^{*}}{L^{*}}\right)\left(\frac{L}{L^{*}}\right)^{\alpha} \exp\left(-\frac{L}{L^{*}}\right). \qquad (10.2)$$

In this expression: ϕ^{*} is a normalization factor that defines the overall density of galaxies (number per cubic Mpc). L^{*} is a characteristic galaxy luminosity. An L^{*} galaxy is a bright galaxy, roughly comparable in luminosity to the Milky Way. A galaxy with $L < 0.1L^{*}$ is a dwarf. ! * α defines the "faint-end slope" of the luminosity function. a is typically negative, implying large numbers of galaxies with low luminosities.

What you should note from Figure 10.7 is that spheroidals (E and S0) dominate for high mass. The disk and irregular galaxies (Sab-Scd and Sd-Irr) play a much more important role at lower masses. And the lowest masses are dominanted by Irr and low surface brightness (LSB) types of galaxies (Moffett et al. 2016a).

This can be taken even further and we can explore how many bulge and spheroidal galaxies there are at a given stellar mass and how many in a disk (Figure 10.8). This comes partially from seeing disk galaxies with a bulge as two distinct components: a bulge (old stars, pressure supported) and a disk (younger stars, rotationally supported). In this clear breakdown, a bulge could be seen as really small spheroidal galaxy surrounded by another!

Breaking every galaxy into spheroidal and disk components (ellipticals have no disk component, S0 very little), the relative fraction of stars in a spheroid and the fraction in disks as a function of stellar mass of the component (in a galaxy or part thereof) is plot in Figure 10.9. What we see is that at almost all masses ($\log(M_{*}) < 11$), stars live predominantely in disks but in the most massive components, spheroids dominate (Moffett et al. 2016b).

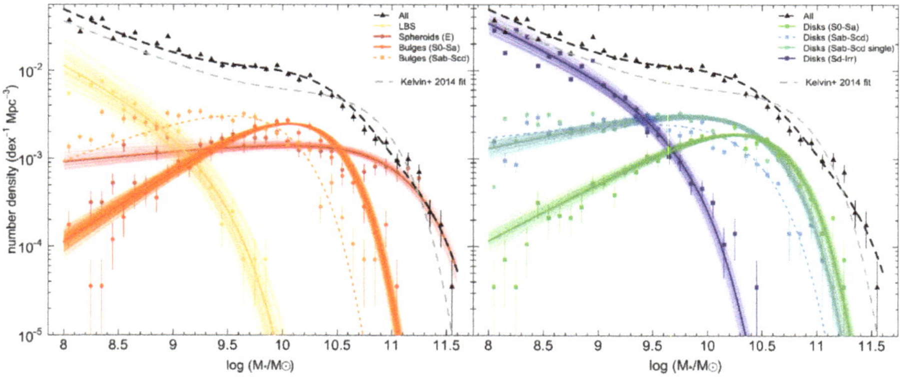

Figure 10.8. Stellar mass functions (the volume density of galaxies as a function of the galaxy's stellar mass) for spheroids (left panel) and disk galaxies (right panel). Fit lines are single Schechter functions. The combined mass function of all components is shown in black, and we also plot the double Schechter total mass function (Kelvin et al. 2014). Figure reproduced from Moffett et al. (2016b). © 2016 The Authors, Published by Oxford University Press on behalf of the Royal Astronomical Society.

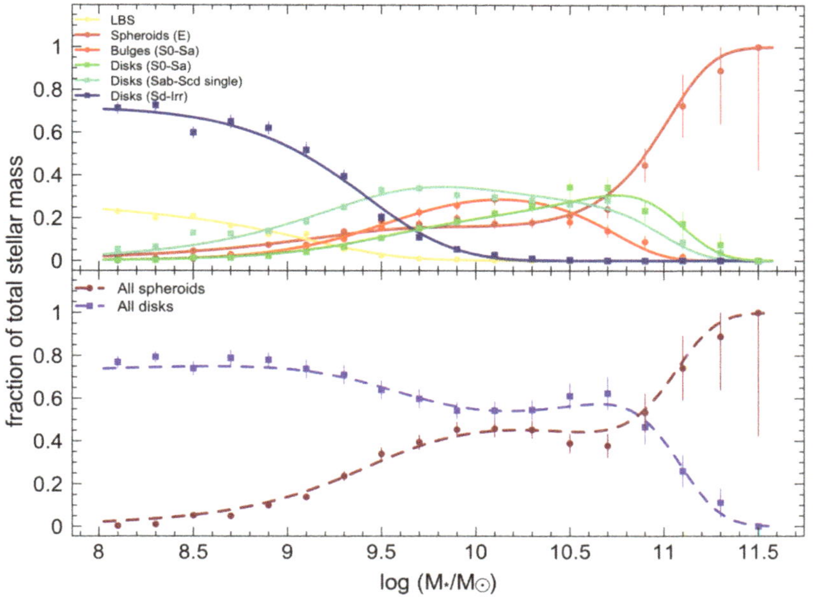

Figure 10.9. Fraction of total stellar mass contributed by each population as a function of stellar mass regime, where points indicate data and lines are derived from the Schechter function fits (Figure 10.8). Due to their indeterminate nature, LBSs are excluded from combined spheroid and disk categories. Figure reproduced from Moffett et al. (2016b). © 2016 The Authors, Published by Oxford University Press on behalf of the Royal Astronomical Society.

10.5.2 Mapping the History of Galaxies

A good way to dissect a galaxy is to map the light it emits across wavelength from the ultraviolet all the way to the sub-mm wavelengths. The measure of how much light is emitted from images ranging from ultraviolet to sub-mm is called the spectral energy density (SED) of an object. This way, the light emitted by young, massive stars, all the much longer-lived stars, the dust and hot gas can be comprehensively modeled. One can do meticulous accounting where ultraviolet photons from stars have been absorbed to account how the dust has the temperature it has. The key parameters for such a model are (a) the stellar mass of the galaxy, (b) the rate it is forming stars, and (c) the mass of dust within the galaxy. This is an analysis you can do on nearby galaxies but also galaxies in the distant (and therefore past) universe.

One can take it even a step further and model *spectra* of galaxies for the relative contributions of ages of galaxies and how much they chemically have evolved, their metallicity.[2] This gives you a measure of how much of the galaxy was made how long ago.

We now have two ways to use the SEDs of galaxies to map out their history over cosmic times: first one can take a census of galaxies in the near (and present) universe

[2] Metallicity in astronomy means the level of non-hydrogen and helium atoms there are in a star. To astronomers, anything not hydrogen or helium is a "metal." Let astronomers mix drinks with caution.

and compare to ever more distant (and hence longer ago) galaxies. Driver et al. (2018) did this using three surveys of galaxies, each progressively further away. In a volume of the universe, how much stellar mass existed? How much star formation was going on? Meanwhile the universe itself is expanding, so that needs to be accounted for. Figure 10.10 shows the density of star formation, stellar mass and the mass of dust in galaxies as inferred from thousands of galaxies in the universe near and far. Density is corrected for the expansion of the universe; these are co-moving (along with the expansion of the universe) volumes. One can spot some clear trends: star formation *density* increased rapidly more than 10 Billion (Gyr) ago and then slowly declined, as the star formation declined. Stellar mass density increases fast initially and grows more steadily over most of the age of the universe. Dust grows first with the star formation and then enters a steady-state, almost as much dust is destroyed as it is produced.

10.5.3 Galaxy Archeology

Meanwhile the second approach is to model the SED of galaxies in the nearby universe to parse apart what the age composition of these galaxies is. This approach is generally called "galaxy archeology." Bellstedt et al. (2020) does this for nearby galaxies in GAMA and finds which *present day* morphological type built up half its stellar mass when (Figure 10.11). The archeology also produces the epoch when each generation of stars in the examined volume was formed so one gets a star formation history at the same time (Figure 10.11, top panel). General trends are that ellipticals formed first, spirals much later and irregulars later still. The overall star formation density aligns very well with what we are seeing in the distant universe (Figure 10.10, top panel) but with the added information with where those earliest formed stars ended up in the present day galaxies.

There is another way to do galactic archeology: resolved stellar populations in other galaxies. As one can image, this is only doable in galaxies close enough to resolve stars in the *other* galaxy. In practice, this is mostly done in the nearest 20 Mpc and with the Hubble Space Telescope. For example, the Panchromatic Hubble Treasury (Dalcanton et al. 2012) program mapped the stellar populations of the nearest spiral galaxy. Comparing the ages of stellar populations in galaxies to what we observe in the distant universe allows us to synthesize a coherent picture in how galaxies evolve.

10.6 Simulated Galaxies

Computer simulations of galaxies are a great tool to understand what processes shaped them and how the physics of stars and star formation relate on the scales of galaxies. The great benefits are that one can speed up the time to include billions of years of evolution, change effects and try again, and modify camera angle on the galaxy that is being studied, all impossible from observations. The great drawbacks of simulations are that they are always limited by their resolution (memory) and much of it is prescriptive, that is within a volume element, there are rules ("prescriptions") of how gas is converted into stars.

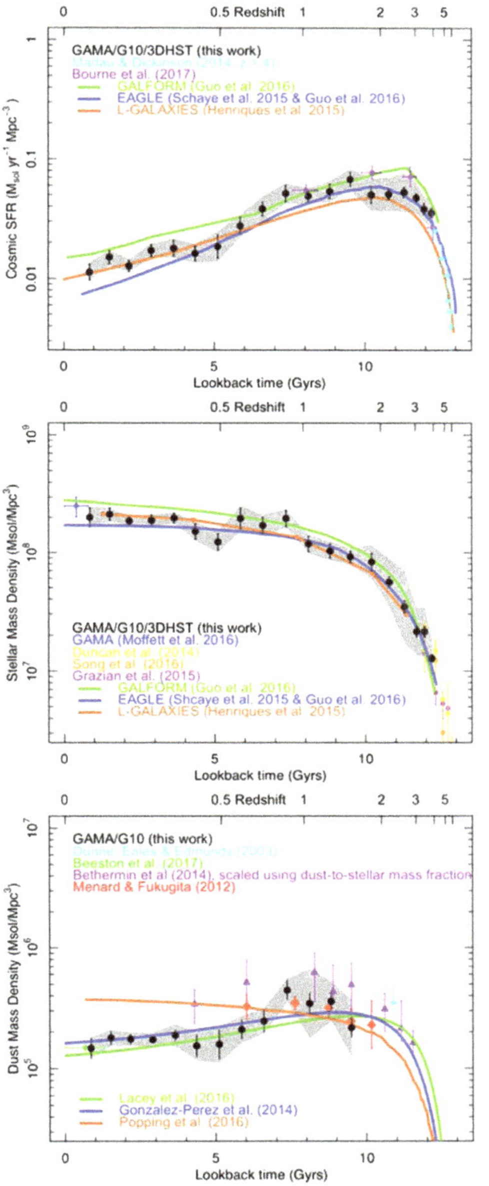

Figure 10.10. Cosmic census of galaxy ingredients. If one looks at ever more distant galaxies and dissects which light is from old and young stars, how much dust intervenes, one can construct the average stellar mass, rate of star formation and dust mass in a cosmic volume. Here, the derived cosmic star formation history (upper panel), stellar-mass build-up (middle panel), and dust-mass build-up (lower panel) in a cosmic volume as a function of lookback time (or redshift). Overlain are model lines. We can see that star formation peaked per volume at 10 billion years ago and have been declining steadily, stellar mass has built up fast and then gradually and dust appeared and mostly stayed at a similar density. Figure reproduced from Driver et al. (2018). © 2017 The Author(s), Published by Oxford University Press on behalf of the Royal Astronomical Society.

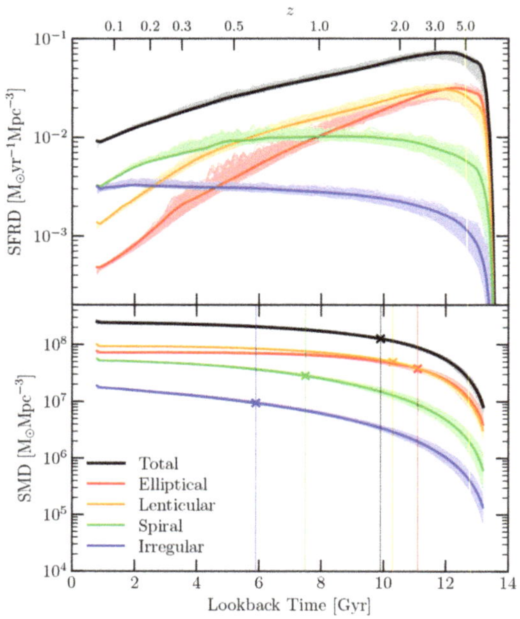

Figure 10.11. Cosmic star formation history for the full sample in black, and the contributions of the individual galaxy morphologies in the colored lines. Bottom panel: cosmic stellar mass density, with the morphologies divided into the same subcategories as in the top panel. For each subpopulation, the epoch at which 50 per cent of the stellar mass is formed is marked with a cross and vertical line. Figure reproduced from Bellstedt et al. (2020). © 2020 The Author(s), Published by Oxford University Press on behalf of the Royal Astronomical Society.

A good starting point for this field is the Toomre & Toomre paper (1972), not because of the simulations used[3] but because they go over the basics of what needs to be considered very well.

Figure 10.12 shows the planes one needs to take into account; the planes of rotation, orbits of galaxies around each other, and the plane of the sky which comes from our viewing angle. Second, their approach is to reproduce well-known objects observed in the sky with a few hundred particles representing the billions of stars in both galaxies (Figures 10.13 and 10.14).

And by-and-large these look like a good job! Of course, there is no gas, ideally we would like a particle for each individual star and track its evolution so one could generate a color image to compare. So in the intervening 50 years, simulations have made steady improvements in their resolution, prescriptions, the inclusions of all observed components, and adding known evolution of stars to the mix. That is a lot to keep track of and that is the reason these are run on supercomputers.

[3] Laughable numbers of particles. You can do litterally do much better on your phone. Your phone can be quite old for that to be true.

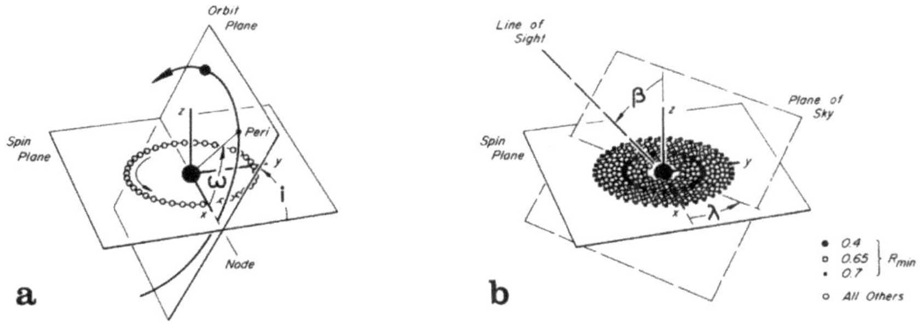

Figure 10.12. An example from Toomre & Toomre on the planes in each simulation: the plane of rotation, the plane of the orbit of the satellite around the central and the plane of the sky defined by our viewing angle. Figure reproduced from Toomre & Toomre (1972).

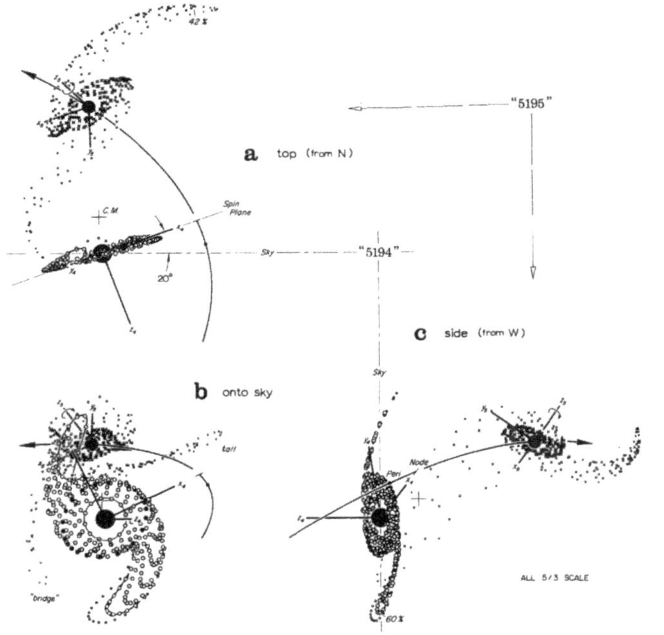

Figure 10.13. The simulations of M51 from Toomre & Toomre. Their simulation already shows tidal effects and it was an excellent first attempt at constraining the orbit of the satellite (NGC5195) to reproduce the spiral arm structure M51 (NGC 5194) is well known for. Figure reproduced from Toomre & Toomre (1972).

10.7 Modern Simulations

Flash forward to today. Computer are much more powerful, surely we understand it all? Not quite. Major successes along the way have been making a disk galaxy and not have it buckle, spiral arm structure, and feedback processes that accurately reproduce the number and luminosities of galaxies we see at different times in the universe.

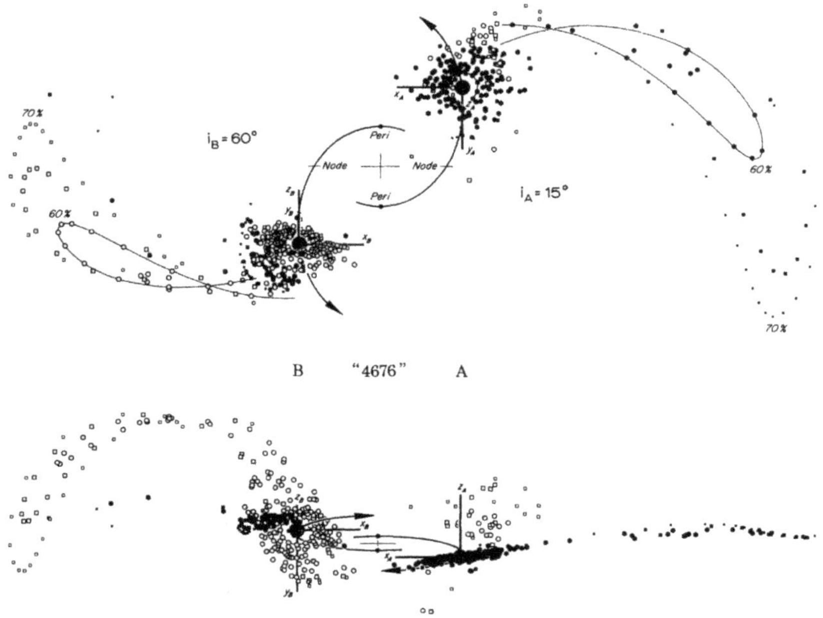

Figure 10.14. The simulations of the "mice" pair of interacting galaxies from Toomre & Toomre. Their simulation nicely shows the tidal arms that are observed. It shows that "peculiar" galaxies such as this could be understood as a brief encounter between two galaxies. Figure reproduced from Toomre & Toomre (1972).

Specifically understanding present-day galaxy appearance (where we have the sharpest images), has been a focus of several cosmological simulations. Once such a simulations is done, the three-dimensional data is reprocessed to resemble observed images as closely as possible: taken from different camera angles, with the starlight put through the filters used for image data. Figure 10.15 illustrates what needs to happen to turn maps of stars and gas into survey images: compute light in the survey filters, add dust where dense gas is present, attenuate and scatter starlight by this dust as a function of observed wavelength and synthesise an image with specific viewing angle. Turning simulations into "observables" is a significant amount of post-production processing.[4]

Then one can analyze these galaxies in the same way as we have been doing all along: the Hubble sequence, galaxy-wide decomposition into bulges and disks, galaxy morphometrics, and a Galaxy Zoo project!

10.7.1 Hubble Sequence

One goal is to reproduce the Hubble sequence (tuning fork) at z=0. If the simulation works as it should, we should be looking at galaxies that fall in this scheme. Using the public Illustris and Sloan Digital Sky Survey images, these can be classified along

[4] A movie making term where the images shot on camera are edited and fixed in "post."

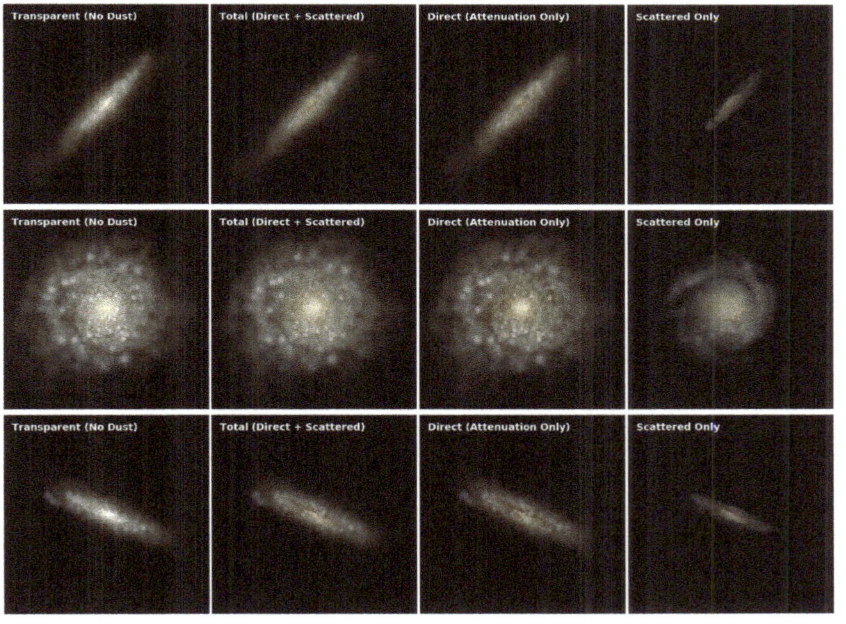

Figure 10.15. Examples of how stellar particle maps are translated into realistic images: First starlight is put through the survey's filters (g, r, and i filters of Pan-STARRS) and then dust (associated with dense gas) is added to redden, attenuate, and scatter the starlight. This gives us a transparent view (no dust), one where all effect are applied (total, what one expects to observe) and relative contributions of each effect, attenuation, and scatter. Camera angles show edge-on (top row), face-on (middle row), and inclined (bottom row). Going from simulated stars and gas to realistic images one can expect to observe is an entire extra set of "post-processing." But once achieved, these can be compared to actual observations directly through any means of quantifying morphology. Figure reproduced from Rodriguez-Gomez et al. (2019). © 2018 The Author(s), Published by Oxford University Press on behalf of the Royal Astronomical Society.

the Hubble tuning fork using machine learning (more about that in Chapter 11). Ordering every galaxy according to their appearance,[5] the stellar mass and size of these galaxies, real and simulated can be compared to each other (Figure 10.16). Not only did Illustris get the appearance right for machine learning to classify, the size–mass relations for each class of object agrees what is observed!

10.7.2 Bulge and Disk Models

Another use of the Illustris galaxies uses the public images, artificially adds them to SDSS to simulate real data and decomposes them into bulge and disk components like the real galaxies. One can implement a single Sérsic profile, leaving the effective

[5] Setting aside if this is the best classification scheme for now, it is the one most astronomers are familiar and comfortable with. Plus 95% of the simulated galaxies fall in one the four main classes.

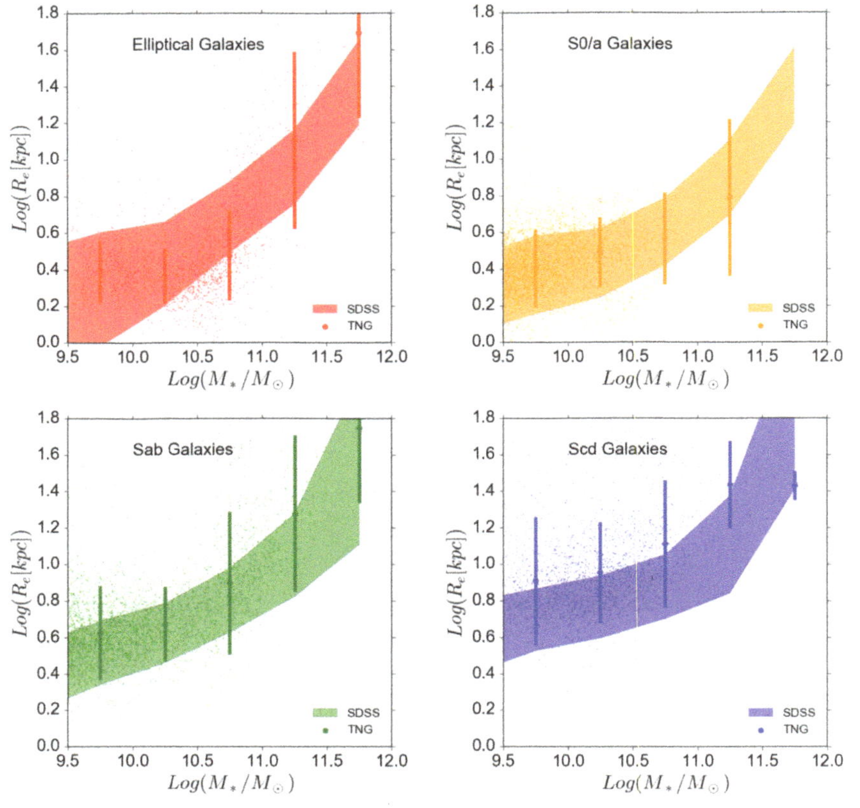

Figure 10.16. Stellar mass–size relation of observed and simulated galaxies divided in four morphological types as labeled. The shaded regions indicate the observed relations from the Sloan Digital Sky Survey. Small dots indicate individual TNG galaxies and the large dots with error bars are median values and scatter. Galaxies in both SDSS and Illustris were classified along the Hubble sequence using a convolutional neural network. Figure reproduced from Huertas-Company et al. (2019). © 2019 The Author(s), Published by Oxford University Press on behalf of the Royal Astronomical Society.

radius and index n free or impose a Bulge+Disk model where one component has to be an $n = 4$ (Bulge) and one $n = 1$ (Disk). The result in Figure 10.17 shows that galaxy-wide measurements do not depend strongly on the choice of descriptor (singe Sérsic or B+D).

10.7.3 Galaxy Morphometrics

Bulge/Disk models or single Sérsic models describe the broad galaxy-wide light distribution. Those seem to (broadly) agree between Illustris and the Sloan Digital Sky Survey images. Looking in more detail, one can use galaxy morphometrics (Chapter 6) to see if both simulated and observed galaxy samples inhabit the same parameter space. Figure 10.18 shows how the morphometric parameter space can be divided up: in disk-dominated, spheroidal galaxies and mergers. In the actual data, many more merging or "peculiar" galaxies are evident compared to the simulations

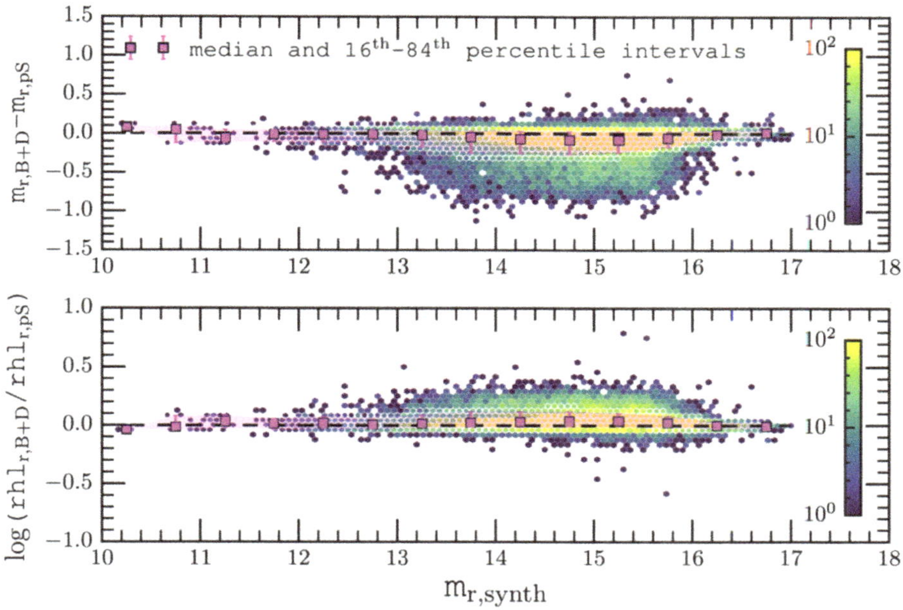

Figure 10.17. A comparison between the catalog values from the Illustris simulation and the measurements using either a Bulge (Sérsic $n = 4$) and Disk (Sérsic $n = 1$) against a Sérsic with n= free profiles on the images. Top panel shows the difference in luminosity and the bottom one the inferrd half-light radius. Figure reproduced from Bottrell et al. (2017). © 2017 The Authors, Published by Oxford University Press on behalf of the Royal Astronomical Society.

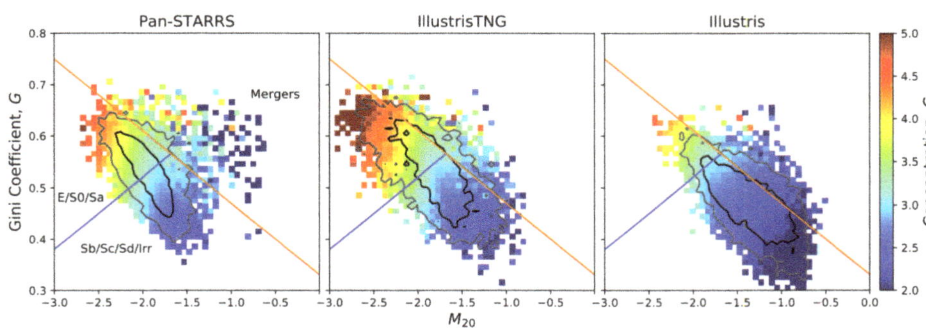

Figure 10.18. The morhometric paramter space (M_{20}–Gini-Concentraion) for the PanStars observations, Illustris and IllustrisTNG (the next generation). Marked are where one would expect disk galaxies, spheroidals (ellipticals and S0) and tidally disturbed mergers. Figure reproduced from Rodriguez-Gomez et al. (2019). © 2018 The Author(s), Published by Oxford University Press on behalf of the Royal Astronomical Society.

(Illustris or IllustrisTNG[6]). A morphometric comparison can reveal in broad terms if the simulations recreate the appearance of the data. In Figure 10.18 we can see that Pan-STARRS and IllustrisTNG agree pretty ok where disk and sphroidal

[6] The Next Generation, more than a few astronomers are fans of star trek. Go figure.

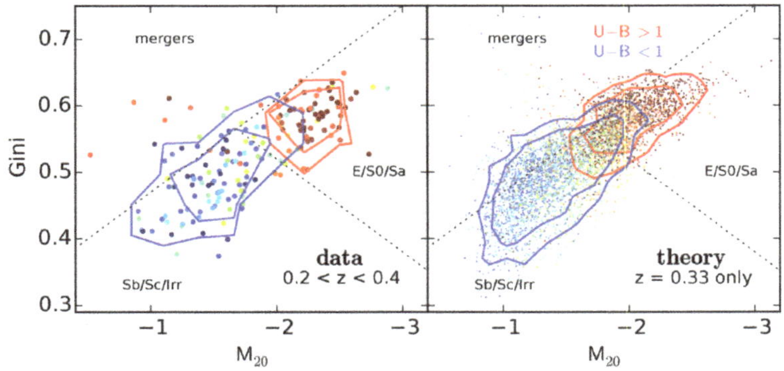

Figure 10.19. The Gini–M_{20} parameter space for Hubble images (left) and Illustris at $z \sim 0.3$ (2 billion years ago). The dashed line shows the divisions between broad Hubble types (disks versus spheroids) and where one expects merging galaxies to be. Compare to Figure 10.18 and notice the following: M_{20} values decline with the x-axis here, the axis is inverted in Figure 10.18. Instead with additional morphological information (concentration), this plot is marked with a color, a proxy for the relative amount of star formation. Blue is star-forming, red is quiescent (not star-forming). There is good general agreement: no star-forming spheroidals, and star-forming disk galaxies. A few mergers that can go either way. Figure reproduced from Snyder et al. (2015). © 2015 The Authors, Published by Oxford University Press on behalf of the Royal Astronomical Society.

galaxies should be and that IllustrisTNG does better than Illustris (TNG makes better looking ellipticals) and where it can do better, mostly not enough merger looking galaxies.

One can look back at higher redshift (z= 0.2–0.4 in this case).[7] Figure 10.19 shows the Gini–M_{20} parameter space for both Illustris galaxies and Hubble data at this $z \sim 0.3$ redshift. The morphological comparison can happen at other redshifts, just be prepared that the usual division into the Hubble tuning fork might not work anymore.

10.7.4 Galaxy Zoo

Another application of the public Illustris data is to show the processed images (made to look like SDSS as much as practical) to Galaxy Zoo volunteers (see Chapter 5). These can be mixed into the hopper of objects being classified to ensure these are treated the same.[8] And Dickinson et al. (2018) found that simulated galaxies and real ones, as viewed by the Galaxy Zoo volunteers: for stellar masses $M_* < 10^{11} M_\odot$, a substantially larger proportion of Illustris galaxies exhibit disk-like morphology or visible substructure compared to the SDSS classifications. They looked at only one question: "Is the galaxy simply smooth and rounded, with no sign of a disk?" (f_{1-F} is the fraction of votes in favor of "not smooth"). Above that mass, the simulated and real galaxies resemble each other much closer (Figure 10.20). With Galaxy Zoo votes, one can identify populations in the morphology parameter

[7] "High" redshift has been a moving marker in astronomy, and in different subfields. Late 1990s? $z = 6$ for Hubble, now? $z = 11$. Radio astronomy? $z > 0.1$.

[8] Human classifiers get squirrelly around fake images. It feels like being fooled.

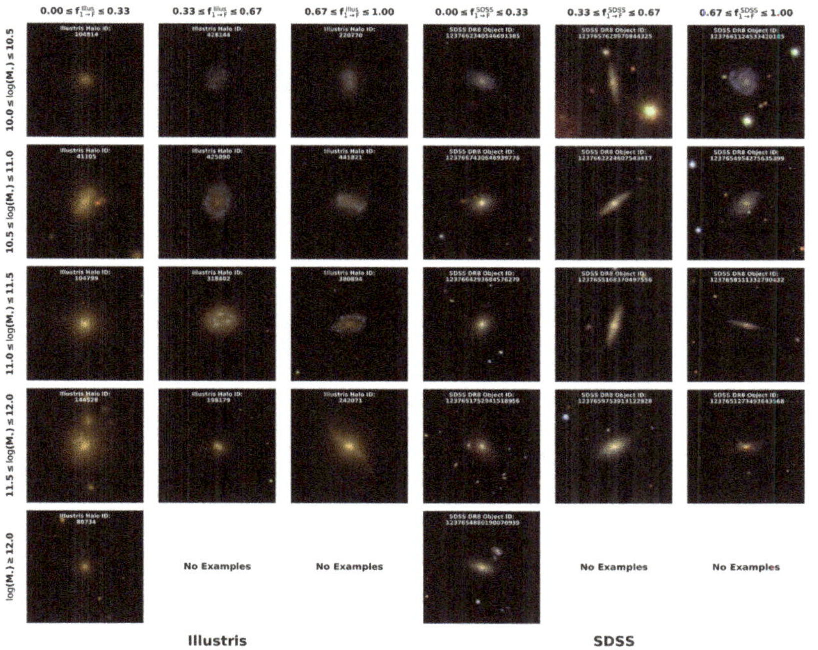

Figure 10.20. Random examples from Illustris (left) and the Sloan Digital Sky Survey as a functon of stellar mass vertical and fraction of votes in favor of features (horizontal). Figure reproduced from Dickinson et al. (2018). © 2018. The American Astronomical Society.

space (>85% of votes for not-smooth) and mass ($M_* > 10^{11} M_\odot$), which go against how we broadly understand galaxy evolution (massive galaxies are ellipticals, so super-smooth) and have a closer look (Figure 10.21). This points to a population of galaxies that almost all Galaxy Zoo classifiers identify as spiral in the SDSS but this population is either missing in the simulated Illustris universe (or classified differently). With simulated galaxies versus real galaxies, the agreements and differences both end up being informative (the next simulation needs an evolutionary path to big spirals).

10.8 Morpho-Kinematics

But why stop here? Increasingly, we have kinematic information on galaxies near and far. Initial classifications of ellipticals into more meaningful categories included kinematic information. Astronomy being a science, these simulations can be used to predict what will be observed in the kinematics of galaxies at different epochs (redshift or lookback times; Thob et al. 2019).

10.9 Assignments

The assignment that pairs with this chapter is "Morphological K-corrections" (jupyter notebook 16). It looks into the effects of looking back with telescopes such as the Hubble Space Telescope and how one can compare to nearby observations.

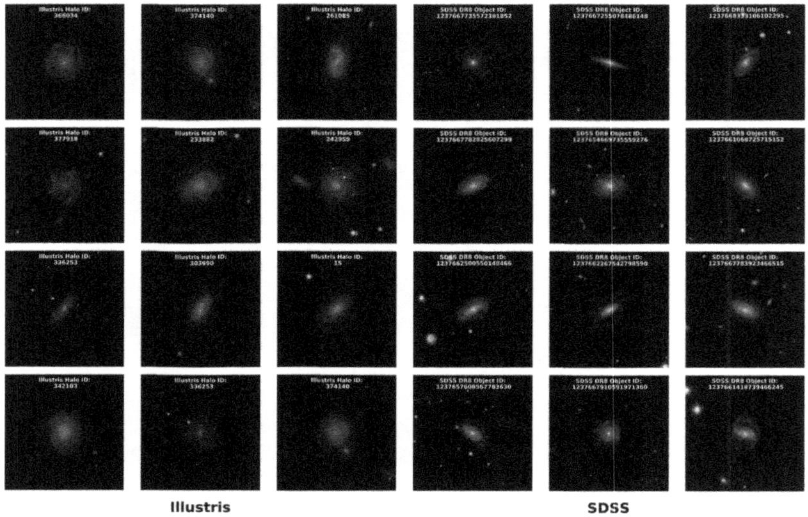

Figure 10.21. Examples from Dickinson where there was a high vote fraction in favor of "not smooth" in SDSS (right) and the closest examples in Illustris (left). Illustris does not produce massive galaxies with disk features. Figure reproduced from Dickinson et al. (2018). © 2018. The American Astronomical Society.

10.10 Further Reading

The book "The Cosmic Evolution of Galaxy Structure" by Christopher Conselice (2020) is an in-depth treatise of what was touched on in this chapter.

References

Bellstedt, S., Robotham, A. S. G., Driver, S. P., et al. 2020, MNRAS, 498, 5581
Bottrell, C., Torrey, P., Simard, L., & Ellison, S. L. 2017, MNRAS, 467, 1033
Buitrago, F., Trujillo, I., Conselice, C. J., & Haußler, B. 2013, MNRAS, 428, 1460
Christodoulou, L., Eminian, C., Loveday, J., et al. 2012, MNRAS, 425, 1527
Conselice, C. J. 2020, The Cosmic Evolution of Galaxy Structure (Bristol: IOP Publishing)
Dalcanton, J. J., Williams, B. F., Lang, D., et al. 2012, ApJS, 200, 18
Dalcanton, J. J., Yoachim, P., & Bernstein, R. A. 2004, ApJ, 608, 189
Dickinson, H., Fortson, L., Lintott, C., et al. 2018, ApJ, 853, 194
Driver, S. P., Andrews, S. K., da Cunha, E., et al. 2018, MNRAS, 475, 2891
Grootes, M. W., Tuffs, R. J., Andrae, E., et al. 2012, in IAU Symp. 284, 352
Holwerda, B. W., Dalcanton, J. J., Radburn-Smith, D., et al. 2012, ApJ, 753, 25
Houghton, R. C. W. 2015, MNRAS, 451, 3427
Huertas-Company, M., Rodriguez-Gomez, V., Nelson, D., et al. 2019, MNRAS, 489, 1859
Kelvin, L. S., Driver, S. P., Robotham, A. S. G., et al. 2014, MNRAS, 439, 1245
Liske, J., Baldry, I. K., Driver, S. P., et al. 2015, MNRAS, 452, 2087
Madau, P., & Dickinson, M. 2014, ARA&A, 52, 415
Masters, K. L., Nichol, R., Bamford, S., et al. 2010, MNRAS, 404, 792
Moffett, A. J., Ingarfield, S. A., Driver, S. P., et al. 2016a, MNRAS, 457, 1308

Moffett, A. J., Lange, R., Driver, S. P., et al. 2016b, MNRAS, 462, 4336
Prescott, M., Baldry, I. K., James, P. A., et al. 2011, MNRAS, 417, 1374
Rodriguez-Gomez, V., Snyder, G. F., Lotz, J. M., et al. 2019, MNRAS, 483, 4140
Rowlands, K., Dunne, L., Maddox, S., et al. 2012, MNRAS, 419, 2545
Scoville, N., Abraham, R. G., Aussel, H., et al. 2007, ApJS, 172, 38
Sheth, K., Elmegreen, D. M., Elmegreen, B. G., et al. 2008, ApJ, 675, 1141
Snyder, G. F., Torrey, P., Lotz, J. M., et al. 2015, MNRAS, 454, 1886
Thob, A. C. R., Crain, R. A., McCarthy, I. G., et al. 2019, MNRAS, 485, 972
Toomre, A., & Toomre, J. 1972, ApJ, 178, 623
Wijesinghe, D. B., Hopkins, A. M., Brough, S., et al. 2012, MNRAS, 423, 3679

Chapter 11

Future Outlook

11.1 Future Observatories

There are many future observatories planned for extra-galactic astronomy. In the optical and near-infrared, the Vera Rubin Observatory, and the Euclid and Roman space telescopes. In the mid-infrared, the James Webb Space Telescope (JWST), probing the restframe wavelength optical at high redshift. Plans abound for near-ultraviolet space missions and the Square Kilometer Array promises to revolutionize radio astronomy once again. There will be images of galaxies. A lot of them.

11.1.1 The Vera C. Rubin Observatory (VRO)

The Vera C. Rubin Observatory will conduct the Legacy Survey of Space and Time (LSST)[1], a survey in the same filters as the Sloan Digital Sky survey but (a) over the Southern Hemisphere, (b) much deeper, and (c) iterated many times to look for things that blink or move in the night's sky.

Deep, uniform images in *ugri* filters will most certainly help galaxy morphology. The era where one needs to actually go out and observe in a broad optical filter is effectively over: this information is available, be it from the Sloan Digital Sky Survey (SDSS; Abolfathi et al. 2018), the Dark Energy Survey (DES; Bechtol et al. 2015), and soon the LSST survey on the Vera Rubin Observatory.

For galaxy morphology studies, the science should include separate measures of bulge and disk, and spiral-arm structure, measures of concentration, asymmetry, and clumpiness as a matter of pipeline processing (Robertson et al. 2017; see Chapter 6). More bespoke measurements will certainly be possible for individual researchers but the sheer amount of galaxies expected to be detected make this data

[1] This was a clever bit of sleuthing since the Rubin Observatory was originally called the Large Synoptic Survey Telescope https://www.lsst.org/

doi:10.1088/2514-3433/ac2c7dch11

a prime candidate for machine learning (ML) techniques which are able to efficiently separate galaxy populations into different morphological classes.

The expectation is that VRO will be producing science-grade observations by the time of publication of this book. An interesting shift will be that the analysis will go to the data as opposed to the other way round: the astronomer uploads scripts to select and process the data to a central repository. The expected data-rate and volume is so large, this cannot easily be stored by individual researchers.

11.1.2 Euclid and Roman

In the intermediate future, two new imaging satellites are set to take over from the venerable Hubble Space Telescope: the European Space Agency's Euclid Telescope (launch in 2022) and the Nancy Roman Space Telescope (launch in the 2030s).

Both of these telescopes are designed to observe much larger parts of the extragalactic sky with resolutions that are close to Hubble's (and no night sky effects), mostly in the near-infrared. Both telescopes are to look for the effects of weak lensing by dark matter on the shapes of galaxies, supernovae going off in the distant universe, and other cosmological probes. The result will be uniform near-infrared (Y, J, H, and Ks) imaging of the sky with plenty of opportunity to dissect the appearance of galaxies in these images.

11.1.3 SKA and Pathfinders

On similar near-future timescales is the use of the Square Kilometer Array (SKA), technologies for which have been developed using pathfinder telescopes and instruments: ASKAP, WSRT/APERTIF, MeerKAT, and JVLA. The concerted effort essentially gave the astronomical community two types of radio observatories on each of the Northern and Southern Hemispheres: a sky-survey interferometer and an interferometric telescope optimized for detailed, pointed observations. The sky-survey interferometer is made possible by the introduction of multiple-dipole cameras at the focal point to the radio telescopes, increasing the effective field-of-view for WSRT and ASKAP well beyond what had been possible. This promises to bring us thousands of reasonably well-resolved galaxies in both radio continuum and H I (about 10 resolution). This offers the opportunity to data-mine the gas appearance of galaxies and just like with Herschel, go beyond unresolved blobs on a massive scale.

At the same time, the rejuvenated VLA (Jansky Very Large Array now) and the MeerKAT radio telescopes have the extent and sensitivity for deep, detailed observations of specific targets (deep fields or individual galaxies). Thus the sky survey can identify targets of interest and these two telescopes can focus on the detail.

SKA is expected to be organized similarly with most of the sky observed in exquisite detail but with components that can go into greater detail and sensitivity.

11.2 Spectro-morphometric

With the common use of Integral Field Units, instruments that render a spectrum for each pixel in the on-sky plane, it is becoming feasible to define morphology components not just on appearance in an image but based on spectral information at the same time.

Large scale IFU surveys are MANGA (Bundy et al. 2015)[2] and SAMI (Allen 2015; Bryant et al. 2015). Using this additional information from the MANGA IFU observations, the Sérsic fit can be improved, implementing limits on the parameter space motivated by the spectral component (e.g., classical bulges are made of old, metal-poor stars; Fischer et al. 2019).

A separate approach can be to isolate and identify the lower surface brightness components with a distinct metal and stellar age. For example, the thick disks of galaxies. Although most of the Milky Way's thick disk's stellar population is considered to have formed *in situ* (Bovy et al. 2012), the structure of the thick disk has been shown by Helmi et al. (2018) to originate mostly from a single mid-sized merger, which brought some 17 kinematically-identified globular clusters into the MW system. The good agreement between the properties of that cluster population (Kruijssen et al. 2019) and those of the thick disk (Haywood et al. 2013; Snaith et al. 2014) imply that *some* of the stellar population *also* came from the merged satellite, see Marsakov et al. (2020) however.

A merger origin, although consistent with properties of the Milky Way disk, need not to hold universally true. VLT MUSE observations of edge-on spiral galaxies have pointed to a lack of lagging or retrograde stellar populations in the thick disk component (Comerón et al. 2019). While a lack of a retrograde moving stellar disk does not need to exclude a merger scenario for the thick disk creation, Comerón et al. (2019) point out the low odds of 1 out of 17 thick disk showing retrograde movement, while 1 in 6 satellite accretions is retrograde. They argue that the prograde kinematics of the thick disk point to a predominantly non-accretion origin of thick disks in spiral galaxies. In contrast, three S0 galaxies of the Fornax cluster *do* show a component of the thick disk originating from a small merger and part of the thick disk forming early and *in situ* (Pinna et al. 2019a, 2019b).

The extra dimension of spectroscopic information for each structural component (Figure 11.1) allows one to construct a history of the stellar population's formation (Figure 11.2). Thus growth-spurts of that component (such as a bump in thick disk growth, one could link to a minor merger) or the spurts, the central nucleus grew in, can become clear.

11.3 Machine Learning

Machine learning covers a wide range algorithms and covering all these is the content of several (thick) books. So we will just go with a short sketch here: these algorithms adapt the relations they use based on the data itself. There are two

[2] Mapping Nearby Galaxies at Apache Point Observatory, an SDSS-IV survey.

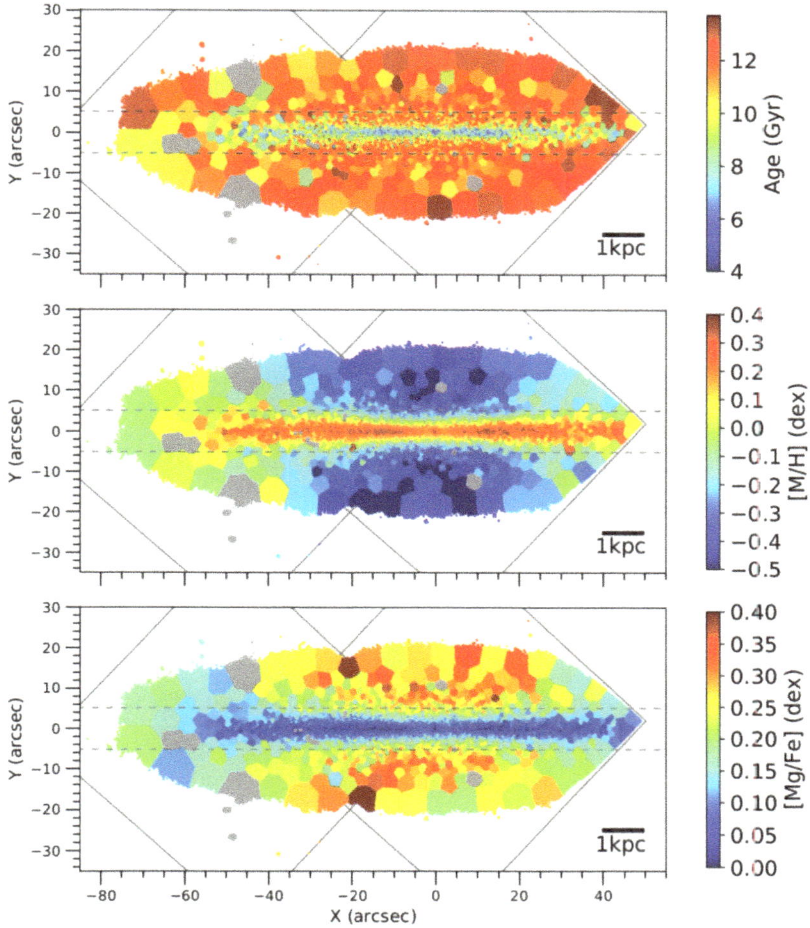

Figure 11.1. The stellar-population maps for FCC 153, an S0 edge-on galaxy in the Fornax cluster. From top to bottom: mean age, total metallicity [M/H], [Mg/Fe] abundance. The discarded bins are plotted in gray, as well as the footprint of the two MUSE instrument pointings. The physical units are indicated by the scale bar on bottom right of each panel. The horizontal dashed lines indicate one scale height above and below the plane. Figure reprinted with permission from Pinna et al. (2019b) © ESO.

general approaches: supervised and unsupervised learning and two outcomes: classification and regression.

Classification is classifying samples into discrete categories e.g., Sa versus Sd etc, which regression is returning a continuous range of values (e.g., predicting the Sérsic effective radius based on an image).

The supervised learning starts with a training sample that has already been classified and learns how to sort that data set based on the labels available. It is then checked with some new data that was not used for training but is of identical quality, the validation sample. Typically the training/validation (or test) samples are 70/30% of all the data there is already a classification for. Only then it is used on unclassified

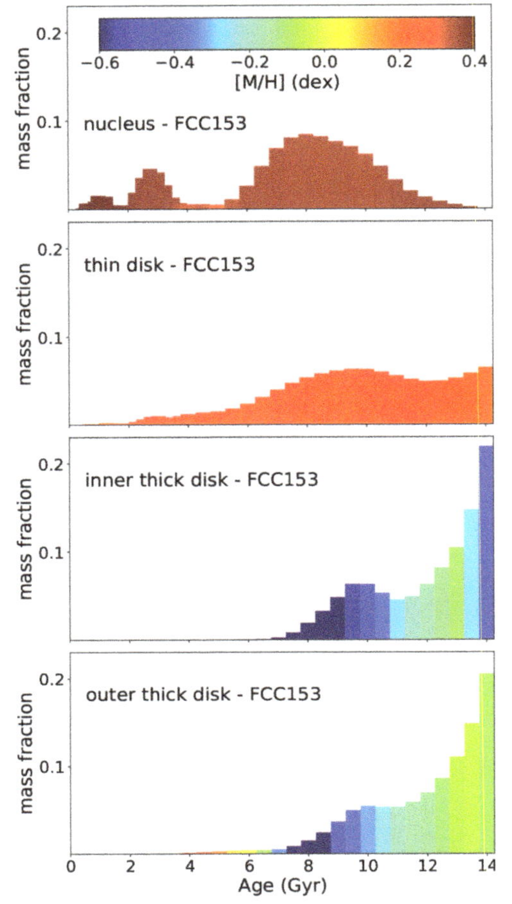

Figure 11.2. The star formation history of the different structural components of FCC 153, from top to bottom: the nucleus, the thin disk, the inner and the outer thick disk. The histogram age bins are color coded according to the weighted average total metallicity ([M/H]) of the specific age bin. The average mass fraction is on the vertical axis, weighted by the mass in each bin and normalized to the mass of the component. Figure adapted with permission from Pinna et al. (2019b) © ESO.

data. The benefit is that—once trained—a supervised ML algorithm will often perform very well, rapidly classifying large data-sets.

For example, I could train on the third reference catalog (de Vaucouleurs et al. 1991) and make a Machine De Vaucouleur. I would train it on the photographs that he originally classified, linking them to the classification he then gave them. And *if I ever would have exactly the same kind of data* (photographic plates of mostly nearby galaxies), I could then use Machine De Vaucouleur to classify them. But I do not! I have CCD imaging, in different filters, maybe from space. Not the same at all! All the photographic plates have been classified already. How to use a ML algorithm trained on one survey and then use it on another is called the *transfer problem*, and it will introduce additional biases and noise in the classification.

The other issue is that supervised ML can only classify what it has been trained on.[3] So if there is a rare case, or simply new information, this will result in misclassifications. These algorithms are very bad at finding new things, just really good at sorting into known categories.

For example, you can train on SDSS images and then expect to use those on VRO/LSST imaging. Same filters but not the same depth. There is additional information the VRO imaging! Low surface brightness structures for example (Chapter 7). So something that is classified as a dwarf spheroidal in SDSS would (more accurately) be classified as a low surface brightness disk galaxy with a small bulge. But those can hardly be found in the SDSS training set (see the section on "Transfer Learning" below).

Unsupervised machine learning is trained on all the data. So data-collection has to be complete (supervised can be done in advance as long as there is a training set). It is typically completely agnostic of any classifications. Those do not exist and definitely are not fed into the algorithm. However, the data is sorted in piles of objects that are "similar" to a certain degree. The trick here is to give the unsupervised ML algorithm an idea of how many piles and how "similar." It also needs a big sample to do this on. The good news is that afterwards, the classification scheme only needs to label each pile, not each object in the data. So we will end up with a stack of objects that later we can label "Sc" etc. And there may even be a small pile that does not fall into the Hubble classification fork, i.e., green spheroidals. Unsupervised machine learning can isolate and help identify new classes or categories of objects!

Golden Globe or "closed envelope" Test

Back in Chapter 1, I mentioned that visual classifications by an expert are still very much a thing (Nair 2010). Such samples of galaxy images with classifications are ideal machine learning training sets! And if one is a good ML scientist, the sample is split into 70/15/15% with 70% used for training, 15% for validation (immediate test), and 15% as a "closed envelope" test of the algorithm.

11.3.1 Catalog Machine Learning

Catalog-based machine learning techniques do not work on the images themselves but the values inferred from the images: color, size, morphometrics (see Chapter 6), or a combination. The trouble here is to offer it a reasonable data set without too much redundancy (because that is weighting that metric twice) but without designing information out.

Each galaxy is then represented by the vector of its measured characteristics (in ML terminology the "feature space," picking the right ones is "feature engineering"). A catalog of these vectors is then the input for a supervised or unsupervised

[3] Not a profound statement eh? yet ... easily forgotten.

machine learning algorithm. This mostly depends if one has classifications for (part) of the catalog (e.g., Galaxy Zoo classifications, see Chapter 5).

Perhaps the use on catalogs will help with the transfer problem. Hopefully the measures of galaxies are calibrated enough to be comparable between surveys. In practice however, it will likely propagate still, just more subtle.

11.3.2 Profile Machine Learning

In between morphometric and full image machine learning, one can train a convolutional neural network (CNN) to predict the pixel profile of galaxies (Smith et al. 2020). Figure 11.3 shows how one can label an image with a profile rather than a single classification number. A CNN is then trained to produce the profile for unseen test images. There is still some offset between the CNN profile and one obtained using an ellipse fit of the image (Figure 11.4) but the method holds potential to rapidly produce surface brightness profiles for millions of images.

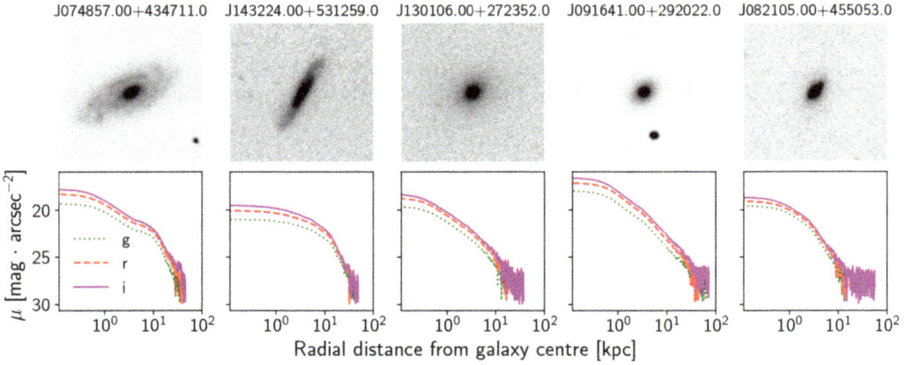

Figure 11.3. A few of the SDSS training images of sample galaxies in the sdss-g band (top row), and corresponding surface brightness "ground truth" profiles (bottom row) with μ as the surface brightness. Instead of fitting ellipses to the top row of images, the images were labeled with the profile below. Figure reproduced from Smith et al. (2020). © 2020 The Author(s), Published by Oxford University Press on behalf of the Royal Astronomical Society.

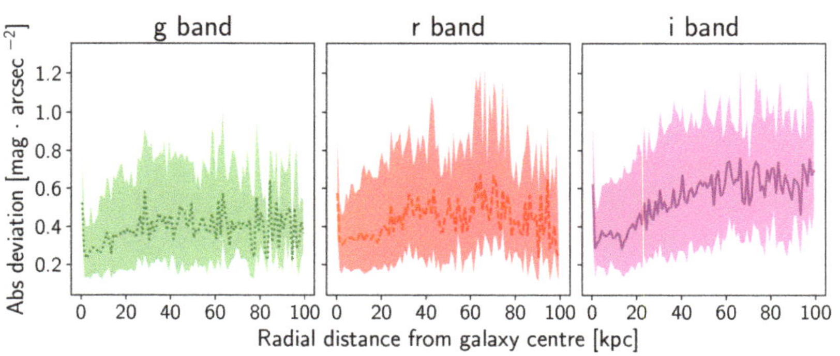

Figure 11.4. The median residual profile between the predicted profiles, using a CNN, and the actual profiles using classical methods (ellipse fit). There is an offset of a few tens of mag arcsec^{-2}. Figure reproduced from Smith et al. (2020). © 2020 The Author(s), Published by Oxford University Press on behalf of the Royal Astronomical Society.

11.3.3 Image Machine Learning

The other approach is to forgo the catalog step and cut out postage stamps of the galaxies and train on those. In practice, any image is a matrix of pixel values and those can be used in machine learning directly. The tacit assumption is that there are clear central positions from which the postage stamp is cut, there is a typical object size to scale the postage stamp, and there are ways to remove nearby objects.

Convolutional neural networks for example are the tool of choice to classify images. This can rely on a lot of work already done by industry (e.g., Google photos, Facebook image recognition etc) so excellent tools already exist. Once again one is limited by what exists in terms of training samples. But for searches of astronomical archives this can be an invaluable tool: once I identify the object of interest, it can suggest objects of similar appearance in the archive.

An example of a CNN schematic is shown in Figure 11.5. Here the postage stamps in color or black and white are fed to the CNN. Each image (an array of values) is convolved with different kernels (red boxes) and then the size is reduced using Max-Pool (selecting out those nodes that carry information) down to a binary choice: is the galaxy merging or not? Note that there are several choices wrapped up in the CNN: what size postage stamp to start with? Color or black and white input? And importantly, what to train for? This CNN will do a good job on merger versus non-merger classification but nothing else.

But one can use a CNN to do a regression too! Tohill et al. (2021) use CNN (deep learning) to predict the morphometrics values, skipping the whole measurement part. Or classify the Hubble sequence but with finer detail and independent typing (Cheng et al. 2021).

Amazingly, the predicted regression value need not be morphology related at all! Wu & Peek (2020) predict the spectrum of SDSS galaxies based on their *ugrizy* images from the Pan-STARRS survey.

Briefly, the method uses a CNN, trained on $56'' \times 56''$ *grizy* image cutouts of galaxies from the Pan-STARRS 3πSteradian Survey DR2, and predicts the galaxies'

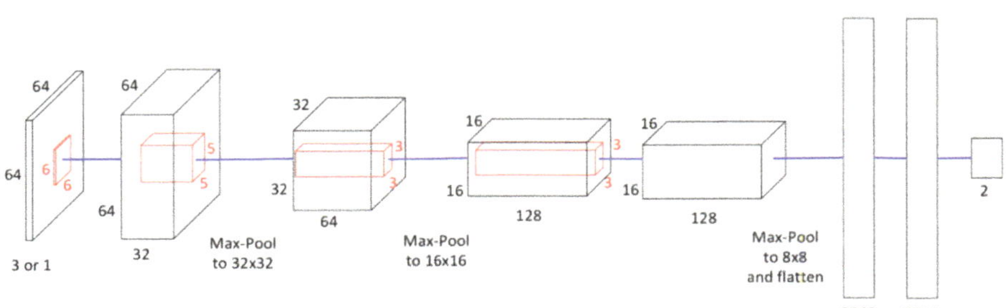

Figure 11.5. The schematic of the convolutional neural network used by Pearson et al. (2019). Starting with postage stamps of the images on the left (64 × 64 pixels in 3 colors or 1 black and white image) going through convolutions (e.g., nodes with edge-detections etc. the size of each kernel is denoted in red) and dimensional reductions (Max-Pool steps) to arrive at a binary classification: merging or not merging. Figure reproduced from Pearson et al. (2019). © 2019 The Author(s), Published by Oxford University Press on behalf of the Royal Astronomical Society.

optical spectra (3″ aperture, SDSS resolution and s/n). Galaxy spectra are normalized, shifted to a common reference frame, and rebinned in 1000 logarithmically spaced elements. A variational autoencoder (VAE) represents each spectrum using six latent variables, which can be robustly "decoded" back into SDSS spectra (see Figure 11.6; Portillo et al. 2020). The CNN prediction is therefore not the whole spectrum but just the six variables that encode most of the information in the spectrum. Wu & Peek (2020) train a CNN to predict galaxies' SDSS spectra solely using image cutouts. The VAE encodes the spectra well and the *grizy* images are enough information to train a CNN on to predict these six values (see Figure 11.7).

11.3.4 Example of Image CNN

Holwerda et al. (2021) use the prediction from Wu & Peek (2020) for a unique object, UGC 2885 ("Rubin's galaxy," Figure 2.10), the largest and most massive, isolated disk galaxy in the local universe (D < 100 Mpc). Big disk galaxies are fairly rare. So the expectation was that this galaxy would be hard to classify using a CNN (an edge-case).

The ML-predicted spectrum (Figure 11.8) could be compared to all existing spectroscopic information that is comparable to an SDSS spectrum of the central region: two archival spectra, one extracted from the VIRUS-P observations of this galaxy, and a new, targeted MMT/Binospec observation (Figure 11.9), all targeting the center of UGC 2885. All this data was obtained or found[4] after the prediction was made. In this sense it was a closed envelope result.[5]

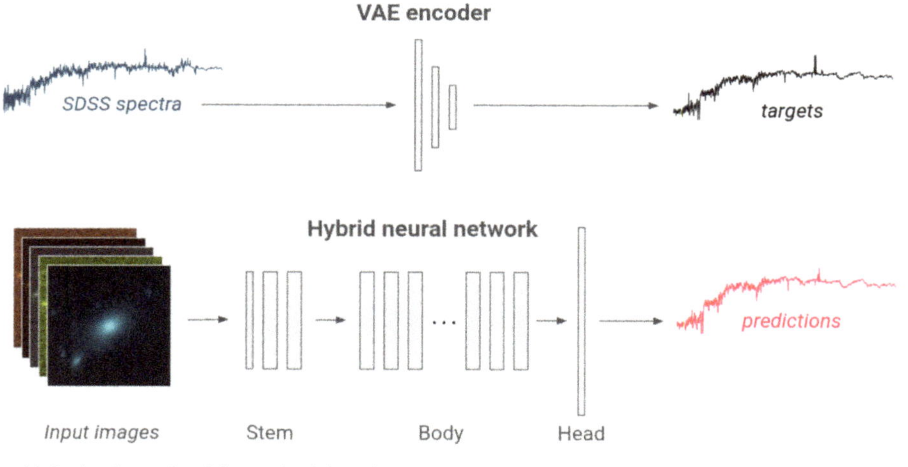

Figure 11.6. A schematic of the methodology in Wu & Peek (2020). A pretrained VAE maps SDSS spectra to six latent variables, which are then used as training targets (labels) for the CNN using the *grizy* galaxy images. The best model comprises a deconvolution stem, resnet-like CNN body, and fully-connected layer head. While the loss function compares targets and predictions in the six-dimensional latent space, we show examples of the decoded spectra for visual comparison. Figure reprinted with permission from Wu & Peek (2020).

[4] Specifically in Bill Keel's paper archive as a printout. Paper lasts.
[5] Or maybe an envelope we found later behind the couch.

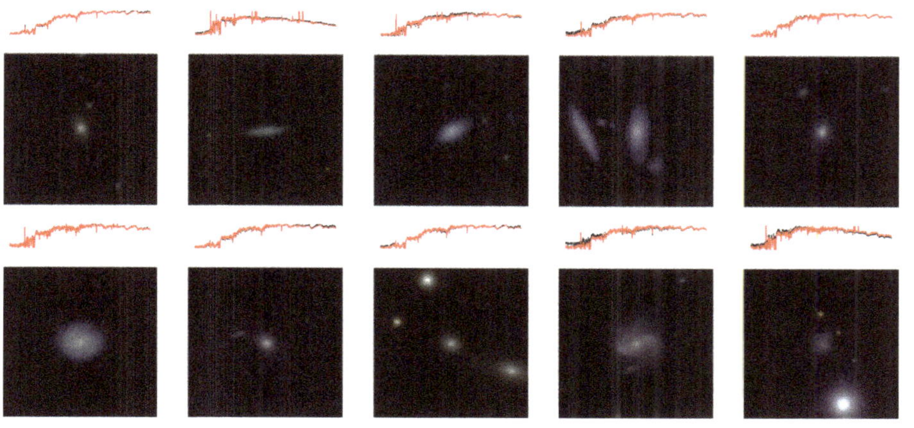

Figure 11.7. Randomly selected examples of Pan-STARRS *grizy* image inputs for the CNN and reconstructed galaxy spectrum outputs. VAE-decoded targets and predictions are shown in black and red, respectively. The size VAE encoders are sufficient to fully capture most SDSS galaxy spectra. Figure reprinted with permission from Wu & Peek (2020).

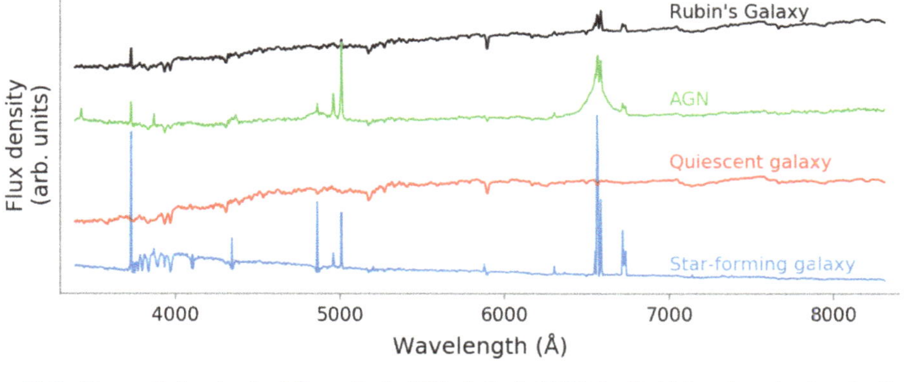

Figure 11.8. The prediction by the ML method of Wu & Peek (2020) for Rubin's galaxy (top) and reference SDSS spectra for an example of an AGN, a quiescent galaxy, and a star-forming galaxy, all shifted to the restframe. The spectral features of Rubin's galaxy fall between those of a quiescent and an AGN-dominated galaxy. Figure reproduced from Holwerda et al. (2021).

Agreement between the ML prediction and all the spectra is qualitatively good (Figure 11.10), though the ML prediction prefers line ratios slightly more toward those of an active galactic nucleus (AGN), compared to archival and VIRUS-P observed values. The MMT/Binospec nuclear spectrum unequivocally shows strong emission lines except Hβ, the ratios of which are consistent with AGN activity.

The CNN trained on *galaxy morphology* got a *spectrum* of the nucleus mostly correct! This is a pretty strong case for the power of CNNs. One can get inferences

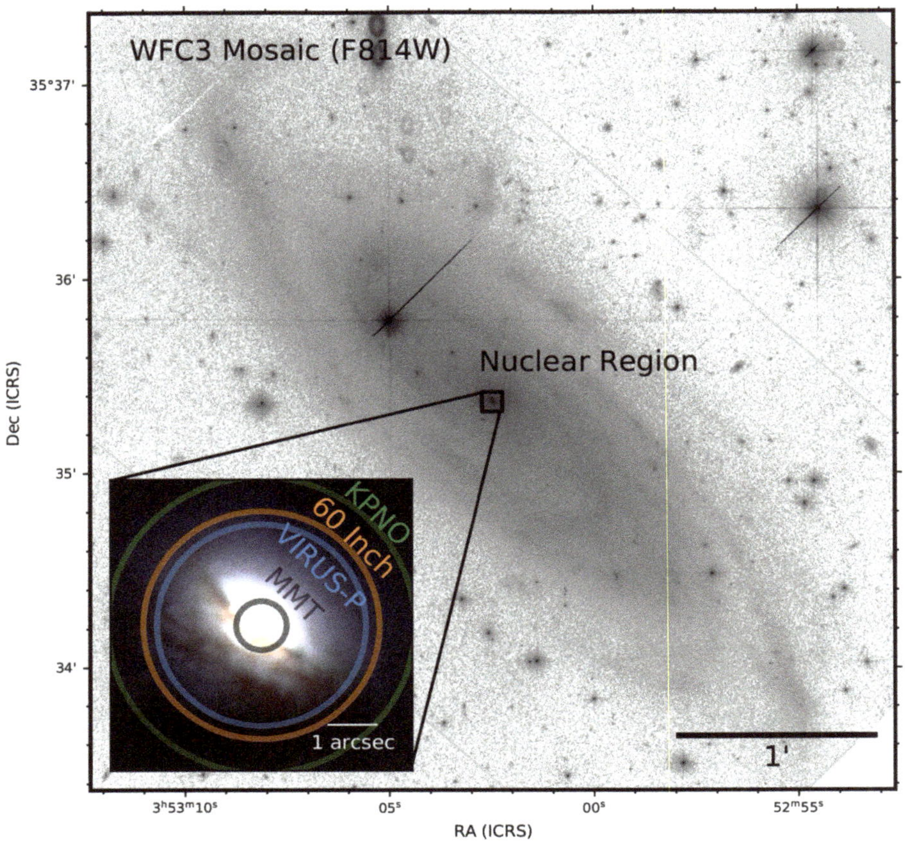

Figure 11.9. The nucleus of Rubin's galaxy from the larger HST mosaic. A clear dustlane partially obscures the central stellar cluster. The inset shows the apertures of the spectroscopic observations are marked using the same color scheme as in Figure 11.10: Mount Lemon 60inch (orange), 2.1 KPNO (green), VIRUS-P (blue), and MMT (white for contrast). Figure reproduced from Holwerda et al. (2021).

for values that are difficult or observationally expensive to obtain and perhaps the links between morphology and other values (e.g., average galaxy metallicity or gas fraction, see Wu 2020 and Wu & Boada 2019).

I should add a warning here: this only works if the training sample is *complete* and *representative*. Biases can sneak in easily. And of course you'll only hear about our successes (reporting bias).

11.3.5 Transfer Learning

Machine or Deep Learning algorithms show great promise for morphological classification of galaxies, mimicking—or even improving—the visual classifications.

However, these algorithms rely on large training samples of already classified galaxies (i.e., with labels). A key question for using ML/DL classifications in future

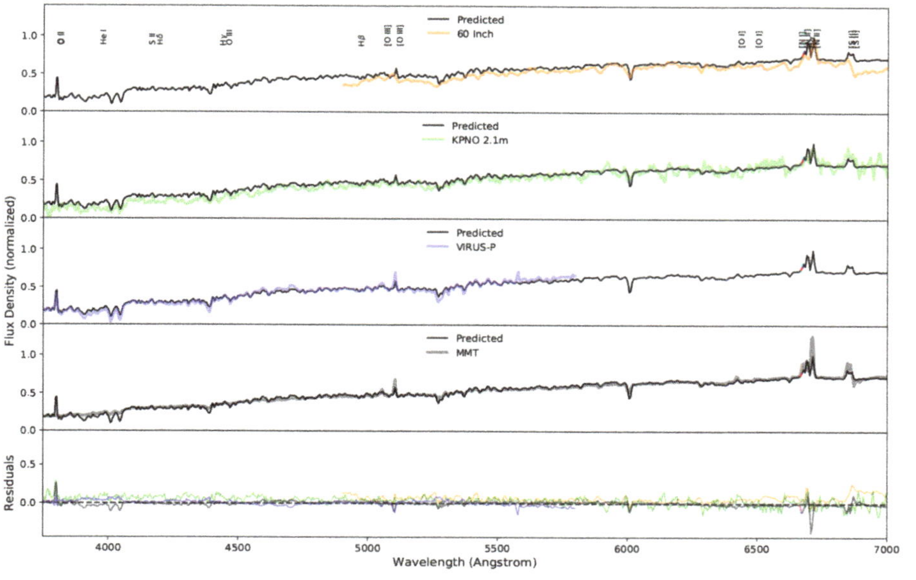

Figure 11.10. The ML-predicted spectrum (black line in top four panels) and the observed spectra, from Keel (1983) using the Mount Lemon 60 inch (top panel, orange), the KPNO 2.1 m telescope (second panel, green), from the VIRUS-P IFU instrument (third panel, blue), and the 2020 MMT Binospec observation (bottom panel, gray). The bottom panel shows the residuals of each spectrum and the ML prediction. Figure reproduced from Holwerda et al. (2021).

surveys is how much of the classifications from an existing survey can be exported to a new survey: how well does it transfer? In machine learning terms this is the "transfer learning" issue. How well is the ML algorithm trained on one survey applicable on a new one?

This can be tested of course, Domínguez Sánchez et al. (2019) compared a deep learning classification trained on the Sloan Digital Sky Survey (SDSS) when applied on the Dark Energy survey (DES), where the filters (colors) are the same but DES is substantially deeper and higher resolution (Figure 11.12).

One can try to apply the SDSS-developed models directly to DES data but that results only in a reasonable global accuracy (~90%), but small completeness and purity values (recall and accuracy). A fast adaptation step, consisting in a further training with a small DES sample of galaxies (a few hundred), is enough for obtaining an accuracy >95% and a significant improvement in the completeness and purity values (Domínguez Sánchez et al. 2019).

The study by Domínguez Sánchez et al. (2019) demonstrates that—once trained with a particular data set—machine learning can quickly adapt to a new survey instrument (e.g., different PSF, seeing, and depth). This is welcome news since it would reduce by almost one order of magnitude the necessary training sample for morphological classification (and therefore the need for much more visual classifications).

It comes down to you cannot blindly apply the ML algorithm from one survey onto another (this didn't really work with morphometrics) but it can be more easily adapted to the new survey. No need to start completely from scratch.

11.3.6 Artificial versus Real Galaxies

Another (future) application of CNNs is to quantitatively compare the galaxy images generated by simulations to those from the Sloan Digital Sky Survey (Zanisi et al. 2021). The metric used to see if the models approximate the real images well is the difference in log-likelihood ratio (LLR). This is the log of the ratio of probability functions: the likelihood this value combination (e.g., R_r and n) appears in the CNN trained on either artificial or real images. The difference in LLR, $\Delta \mathrm{LLR} = \mathrm{LLR}_{\mathrm{sim}} - \mathrm{LLR}_{\mathrm{SDSS}}$ is an indication how close together the images are. Figure 11.11 shows a comparison between three suites of artificial galaxies, Illustris, The Next Generation[6] 50 and 100. Not only can one compare if the distributions of galaxy-wide morphology characteristics (Sérsic index and effective radius) are reproduced well by the simulations but the CNN are sensitive to small-scale features. Figure 11.11 shows that for example star-forming, bigger galaxies are better described than small, quiescent galaxies, a guidance for the next effort in galaxy simulations.

11.3.7 Machine-assisted Citizen Science

Although visual inspection of large observational surveys is time-intensive, the advent of citizen science like Galaxy Zoo has revolutionized its use on surveys (e.g., Lintott et al. 2011; Simmons et al. 2017; Willett et al. 2017). Galaxy Zoo has used more than a million citizen-science volunteers to classify large contemporary surveys, like the Sloan Digital Sky Survey (SDSS), the Kilo-Degree Survey (KiDS), and Hubble Space Telescope Legacy Surveys and has provided the benchmark against which the accuracy of automated techniques can easily be compared (e.g., Huertas-Company et al. 2015; Dieleman et al. 2015; Beck et al. 2018; Walmsley et al. 2019; Ma et al. 2019).

A next logical step is to have a machine algorithm learn with the citizen science effort. As soon as a certain galaxy is classified, similar enough (in a parameter space of choice) galaxies are given the classification. A real-life example is when a photo collection service asks you about whether this person in the photo is indeed your loved one and suddenly many more pictures with the same person in it are also tagged. We saw what happens in Figure 5.9 and Chapter 5: aided with a machine learning mechanism, the citizen science classify a sample much faster, yet at the same accuracy.

11.3.8 Morphometrics ML

Machine Learning can be used to either predict values for the Morphometrics we saw earlier in Chapter 6 or they can use morphometrics catalogs as their input. For example, on the Hubble images from the CANDELS project, one can compute their

[6] Unsurprisingly, a lot of astronomers are fans of Star Trek.

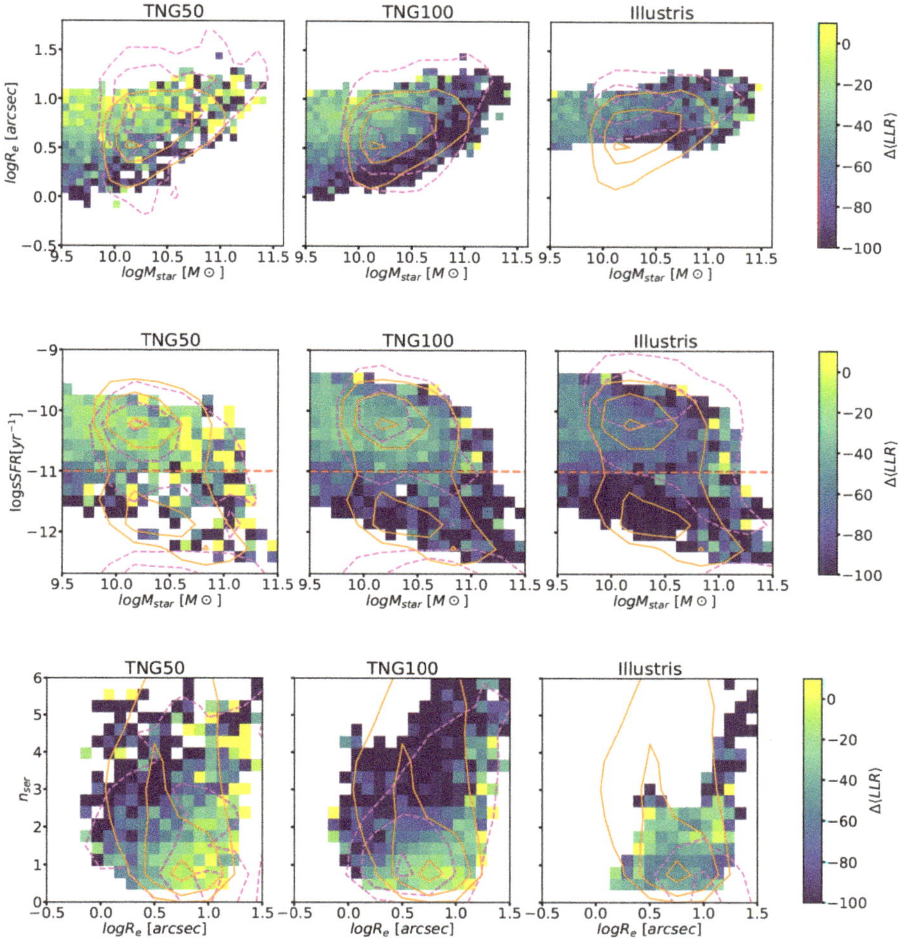

Figure 11.11. The size-mass relation (top panel), sSFR – $M_{\rm star}$ relation (middle panel) and the Sérsic index-size relation (bottom panel) for three simulations. The color code is the difference between the mean LLR of each simulation and the mean LLR of SDSS at each point of the scaling relations. A brighter color indicates a better agreement with SDSS. In the middle panel the red dashed line denotes the sSFR threshold that separates star-forming (log(sSFR) > 11) and quiescent(log(sSFR) < −11) galaxies. The orange solid contours the 10th, 50th, and 90th percentiles of the 2D distributions for real SDSS galaxies (above the completeness threshold of $M_{\rm star} = 10^{10} M_{\odot}$). Dashed magenta contours are the same for each simulation. Quenched (low sSFR), concentrated (high n), and small (low $R_{\rm e}$) galaxies are the ones with the lowest ΔLLR, and so their fine stellar morphology in the simulations substantially disagrees with observations. Figure adapted from Zanisi et al. (2021), Copyright © 2020, Oxford University Press.

Concentration and Asymmetry but one could also train a convolutional neural network to predict these values based on their cutout images (Tohill et al. 2021). With a CNN, the predicted values agree well with what's computed (see Figure 11.13). The benefit of a CNN is that is can be more robust against redshift effects.

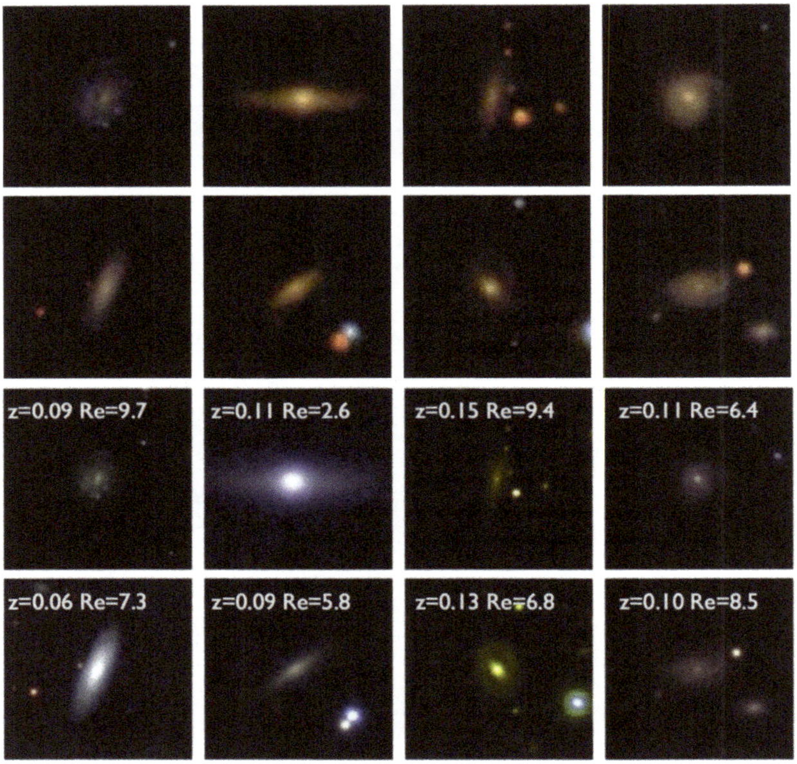

Figure 11.12. Top two rows: six SDSS galaxies, bottom two rows: the same galaxies in DES. The galaxies were randomly selected from the common sample of the two surveys, with the only requirement of having high probability of being disk, edge-on or barred galaxies. The higher quality of DES images reveals with higher detail some galaxy features, such as bulge component or spiral arms. Figure reproduced from Domínguez Sánchez et al. (2019).

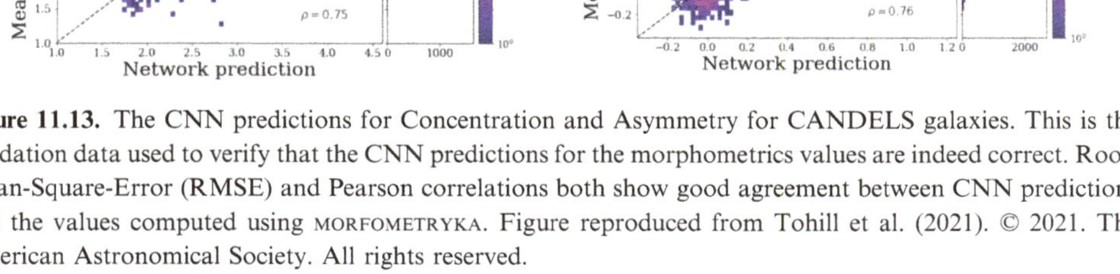

Figure 11.13. The CNN predictions for Concentration and Asymmetry for CANDELS galaxies. This is the validation data used to verify that the CNN predictions for the morphometrics values are indeed correct. Root-Mean-Square-Error (RMSE) and Pearson correlations both show good agreement between CNN predictions and the values computed using MORFOMETRYKA. Figure reproduced from Tohill et al. (2021).

Alternatively, a morphometrics space can be an excellent catalog to train machine learning algorithms on. Each object is in essence a vector in feature space. Figure 11.14 shows where morphometric weight is in a Self Organizing Map (SOM), an unsupervised learning algorithm that sorts a multidimensional feature space onto a single map. This feature space (from Pearson et al. 2019) maps certain groupings as visible in Figure 11.14. The issue with unsupervised learning then is how to label each grouping of nodes on the SOM.

Figure 11.15 shows a first attempt at labeling the features in an unsupervised SOM by using votes from the Galaxy Zoo. For each object, one determines a feature that "won" i.e., which number of spiral arms got the most votes. This illustrates how tricky this sort of classification is. Does the feature space have enough information in it to even discriminate between 2 and 3 spiral arms? Was it trained long enough? Are there even areas that correspond to 4-armed spiral galaxies? And as we have seen in Chapter 5, how well can one even determine the number of spiral arms from a series of votes?

11.3.9 Eigengalaxies

Another technique for morphology classification is Principle Component Analysis (PCA). In principle, this is taking a non-orthogonal space of a sample of vectors (be

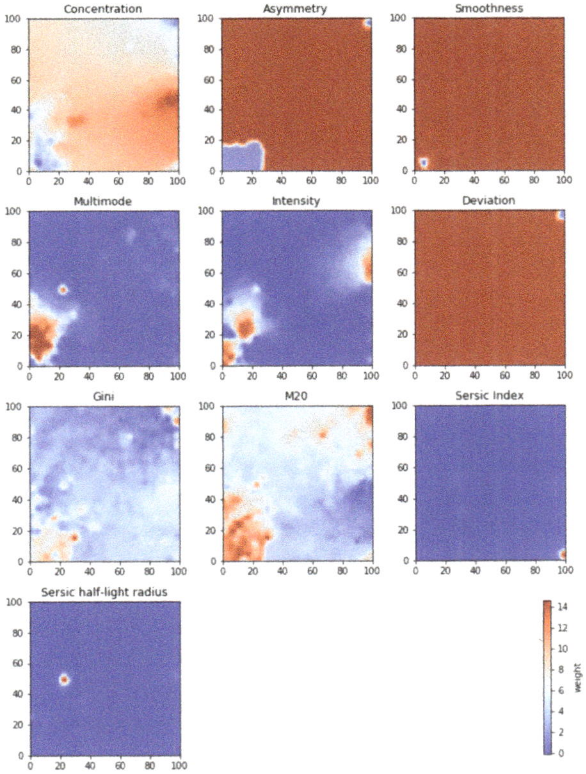

Figure 11.14. The features (CAS, MID, G-M20, and Sérsic) mapped onto the SOM trained on them. Figure reproduced from Smith et al. (2020). © 2020 The Author(s), Published by Oxford University Press on behalf of the Royal Astronomical Society.

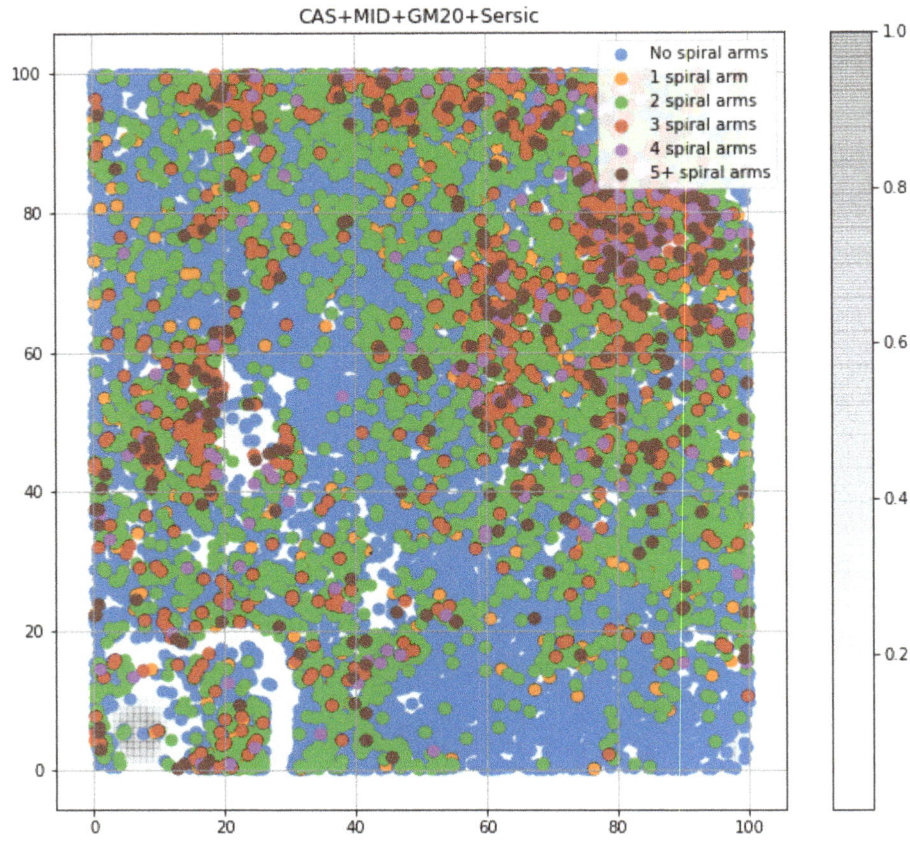

Figure 11.15. The distributions over the SOM instance in Figure 11.14 of the number of arms as determined by the Galaxy Zoo voting. The majority vote for each galaxy for 1, 2, 3, 4, or 5 or more spiral arms. Figure reproduced from Smith et al. (2020). © 2020 The Author(s), Published by Oxford University Press on behalf of the Royal Astronomical Society.

it morphometric catalog entries or simply the image-of-the-galaxy-as-a-vector) and taking the eigenvectors,[7] the orthogonal components describing the sample space. It is a minimalization process: how many independent vectors describe the data? And how good is good enough of a description? In Chapter 6 we saw how the morphometrics space could be orthogonalized in the ZEST system (Scarlata et al. 2007).

One can also simply take out the intermediate step of morphometrics measures and use PCA on the images themselves and calculate "eigengalaxies" (Uzeirbegovic et al. 2020), the principal components of galaxy appearance. The next step is deciding how much (or little) variance from the best description with N principal components you are prepared to live with. Figure 11.16 shows the number of components of the PCA and how much variance with the sample is explained by the PCA model. If 85% of variance reproduced would be good enough, only two

[7] This term may ring a bell if you had Quantum Mechanics recently.

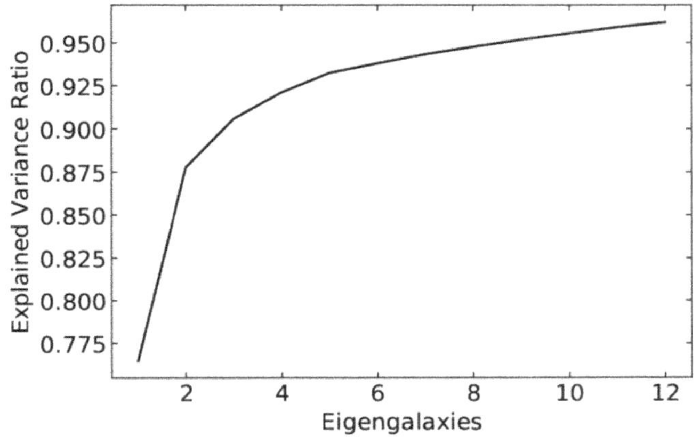

Figure 11.16. Cumulative explained variance illustrating how much variance each additional feature accounts for. In PCA successive components account for less variance than the previous component. In this instance, only two eigengalaxies are required to account for some 85% of explained variance yet 12 eigengalaxies are required to account for 96% of explained variance. Figure reproduced from Uzeirbegovic et al. (2020). © The Author(s) 2020, Published by Oxford University Press on behalf of the Royal Astronomical Society.

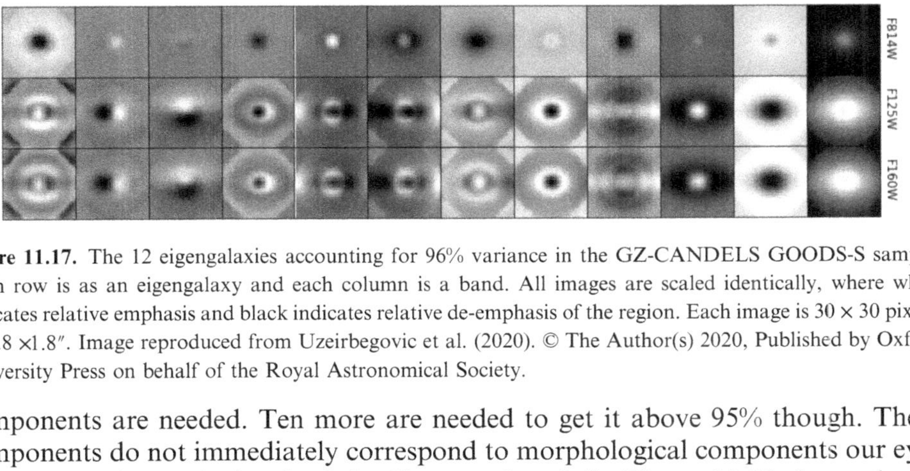

Figure 11.17. The 12 eigengalaxies accounting for 96% variance in the GZ-CANDELS GOODS-S sample. Each row is as an eigengalaxy and each column is a band. All images are scaled identically, where white indicates relative emphasis and black indicates relative de-emphasis of the region. Each image is 30 × 30 pixels, or 1.8 ×1.8″. Image reproduced from Uzeirbegovic et al. (2020). © The Author(s) 2020, Published by Oxford University Press on behalf of the Royal Astronomical Society.

components are needed. Ten more are needed to get it above 95% though. These components do not immediately correspond to morphological components our eyes have assigned to galaxies (back in Chapters 1 and 5). Figure 11.17 shows the 12 components for each filter considered.

Some of the PCA images correspond to typical morphological distinctions (e.g., PCA3 corresponds to a resolved disk? y/n) but the PCA space is best used to describe as a vector space to look for similar populations. Figure 11.18 shows four sub-categories that can be identified as close to an "exemplar" galaxy. Because the galaxies are now converted in a 12-dimensional space, once can take the PCA vector of the exemplar galaxy and sort the full sample by distance from that example galaxy and find the next n galaxies that are closest to it in PCA space. Figure 11.18 illustrates it with four similarity groupings for spirals, edge-on spirals, ellipticals, and recent mergers. The issue is at what point is a galaxy too far separated in 12-dimensional space to no longer be considered "similar"?

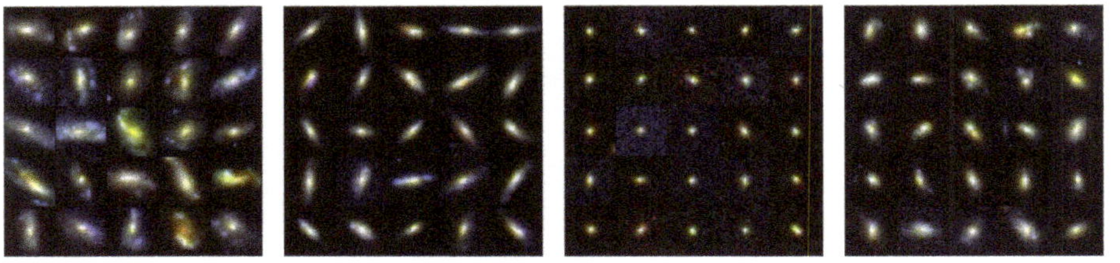

Figure 11.18. Examples of similarity searches in PCA space. The exemplar galaxy is given in the top left, followed by 24 of its nearest neighbors (closest cousins) by distance in the 12-dimensional PCA eigengalaxy space. These show the searches for spirals, edge-on spirals, ellipticals and recent mergers from left to right respectively. Figure reproduced from Uzeirbegovic et al. (2020). © The Author(s) 2020, Published by Oxford University Press on behalf of the Royal Astronomical Society.

The PCA "eigengalaxies" approach is a great way to construct sample of similar galaxies. Once constructed, such a space can be used to find clusters of similar galaxies using machine learning techniques (e.g., K-clustering or similar; Uzeirbegovic et al. 2020). Finding galaxies that are similar to each other can also make a citizen science effort much more efficient (see Chapter 5 on machine-assisted citizen science). This is the same principle as photo services tagging people in pictures after some training.

11.4 Outlook

This is a rapidly developing field in astronomy, in a large part because we expect so much data from the new observatories that checking by eye or even a citizen science approach may be completely impossible. Some of the references here may give you ideas for your next project. Does it involve classifying or regression? Is there a labeled data set? Big enough, spanning enough feature space? Is your data set balanced or are you after rare objects?

References

Abolfathi, B., Aguado, D. S., Aguilar, G., et al. 2018, ApJS, 235, 42

Allen, J. T. 2015, in IAU Symp. 309, Galaxies in 3D across the Universe, 109

Bechtol, K., Drlica-Wagner, A., Balbinot, E., et al. 2015, ApJ, 807, 50

Beck, M. R., Scarlata, C., Fortson, L. F., et al. 2018, MNRAS, 476, 5516

Bovy, J., Rix, H.-W., Hogg, D. W., et al. 2012, ApJ, 755, 115

Bryant, J. J., Owers, M. S., Robotham, A. S. G., et al. 2015, MNRAS, 447, 2857

Bundy, K., Bershady, M. A., Law, D. R., et al. 2015, ApJ, 798, 7

Cheng, T.-Y., Huertas-Company, M., Conselice, C. J., et al. 2021, MNRAS, 503, 4446

Comerón, S., Salo, H., Knapen, J. H., & Peletier, R. F. 2019, A&A, 623, A89

de Vaucouleurs, G., de Vaucouleurs, A., Corwin,, H. G., et al. 1991, Third Reference Catalogue of Bright Galaxies, Vol. 1–3 (Berlin: Springer)

Dieleman, S., Willett, K. W., & Dambre, J. 2015, MNRAS, 450, 1441

Domínguez Sánchez, H., Huertas-Company, M., Bernardi, M., et al. 2019, MNRAS, 484, 93

Fischer, J. L., Domınguez Sanchez, H., & Bernardi, M. 2019, MNRAS, 483, 2057

Haywood, M., Di Matteo, P., Lehnert, M. D., Katz, D., & Gomez, A. 2013, A&A, 560, A109

Helmi, A., Babusiaux, C., Koppelman, H. H., et al. 2018, Natur, 563, 7729
Holwerda, B. W., Wu, J. F., Keel, W. C., et al. 2021, ApJ, 914, 142
Huertas-Company, M., Pérez-González, P. G., Mei, S., et al. 2015, ApJ, 809, 95
Keel, W. C. 1983, ApJ, 269, 466
Kruijssen, J. M. D., Pfeffer, J. L., Reina-Campos, M., Crain, R. A., & Bastian, N. 2019, MNRAS, 486, 3180
Lintott, C., Schawinski, K., Bamford, S., et al. 2011, MNRAS, 410, 166
Ma, Y.-B., Cao, S., Zhang, J., et al. 2019, EPJC, 79, 121
Marsakov, V. A., Koval', V. V., Gozha, M. L., et al. 2020, AstBu, 75, 21
Nair, P. B., & Abraham, R. G. 2010, ApJS, 186, 427
Pearson, J., Li, N., Dye, S., et al. 2019a, MNRAS, 488, 991
Pearson, W. J., Wang, L., Alpaslan, M., et al. 2019b, A&A, 631, A51
Pinna, F., Falcon-Barroso, J., Martig, M., et al. 2019a, A&A, 625, A95
Pinna, F., Falcon-Barroso, J., Martig, M., et al. 2019b, A&A, 623, A19
Portillo, S. K. N., Parejko, J. K., Vergara, J. R., & Connolly, A. J. 2020, AJ, 160, 45
Robertson, B. E., Banerji, M., Cooper, M. C., et al. 2017, arXiv:1708.01617
Scarlata, C., Carollo, C. M., Lilly, S., et al. 2007, ApJS, 172, 406
Simmons, B. D., Lintott, C., Willett, K. W., et al. 2017, MNRAS, 464, 4420
Smith, R. J., Collier, W. P., Ozaki, S., & Lucey, J. R. 2020, MNRAS, 493, L33
Snaith, O. N., Haywood, M., Di Matteo, P., et al. 2014, ApJL, 781, L31
Tohill, C., Ferreira, L., Conselice, C. J., Bamford, S. P., & Ferrari, F. 2021, ApJ, 916, 4
Uzeirbegovic, E., Geach, J. E., & Kaviraj, S. 2020, MNRAS, 498, 4021
Walmsley, M., Ferguson, A. M. N., Mann, R. G., & Lintott, C. J. 2019, MNRAS, 483, 2968
Willett, K. W., Galloway, M. A., Bamford, S. P., et al. 2017, MNRAS, 464, 4176
Wu, J. F. 2020, ApJ, 900, 142
Wu, J. F., & Boada, S. 2019, MNRAS, 484, 4683
Wu, J. F., & Peek, J. E. G. 2020, arXiv:2009.12318
Zanisi, L., Huertas-Company, M., Lanusse, F., et al. 2021, MNRAS, 501, 4359

www.ingramcontent.com/pod-product-compliance
Ingram Content Group UK Ltd.
Pitfield, Milton Keynes, MK11 3LW, UK
UKHW061858100926
8556IPUK00008B/914

9 780750 334976